W9-AST-750

SEA SQUIRTS OF THE ATLANTIC CONTINENTAL SHELF FROM MAINE TO TEXAS

Living specimens of *Boltenia ovifera* dropped from the dredge on the deck of the *Albatross IV*, while cruising ten miles south of Mt. Desert Island in the Gulf of Maine.

Sea Squirts of the Atlantic Continental Shelf from Maine to Texas

Harold H. Plough

Drawings by Mareile Fenner

The Johns Hopkins University Press
Baltimore, Maryland

To Dr. Roland L. Wigley, who stimulated an
old Ascidian taxonomist, with genetic interests,
to study again the distribution of species of
Ascidiacea on the Atlantic continental shelf.

Manufactured in the United States of America

The Johns Hopkins University Press, Baltimore, Maryland 21218
The Johns Hopkins Press Ltd., London

Library of Congress Catalog Card Number 76–47388
ISBN 0–8018–1687–4

Library of Congress Cataloging in Publication data
will be found on the last printed page of this book.

Publication of this book has been assisted
by a grant from the Charles E. Merrill Trust.

Contents

Preface ... vii
I. The Ascidian Structural Plan ... 1
II. Distribution of the Species of *Ascidiacea* in the Gulf of Maine ... 4
III. The Biogeographical Regions on the Continental Shelf and Distribution of Ascidian Species ... 9
IV. Families of *Ascidiacea* Found on the Continental Shelf ... 16
V. Modifications of the Branchial Sac of Evolutionary Significance ... 47
VI. The Species of *Ascidiacea* from Maine to Texas ... 53
VII. Distribution in Relation to Continental Drift ... 94
VIII. The Ascidian Larva and the Beginning of the Chordate Line ... 103
Appendix List of Species of *Ascidiacea* Found on the Atlantic Continental Shelf of the United States, 1968–1974 ... 109
References ... 113
Index ... 115

FIGURES

1. *Ecteinascidia tortugensis*, adult ... 1
2. *Ecteinascidia tortugensis*, development; vegetative cycle ... 2
3. Distribution in Gulf of Maine ... 6
4. *Albatross IV* benthos-collecting cruise of 1968 ... 7
5. Collecting regions off the eastern United States ... 10
6. Off Sapelo Island, Georgia ... 11
7. Florida West Coast ... 13
8a. Location of Apalachee Bay, Florida ... 14
8b. Apalachee Bay, Florida ... 14
9. Padre Island, Texas ... 15
10. Distribution of fifteen species ... 15
11. *Eudistoma olivaceum* ... 20
12. *Perophora viridis* and *P. bermudensis* ... 21
13. *Ecteinascidia turbinata* and *E. tortugensis* ... 22
14. *Corella borealis*; *Chelyosoma macleayanum*; *Rhodosoma wigleii* ... 39
15. *Ascidia prunum* ... 40
16. *Styela partita* ... 41
17. *Styela plicata* ... 42
18. *Polycarpa fibrosa* ... 43
19. *Cnemidocarpa mortenseni*; *C. mollis* ... 44
20. *Dendrodoa carnea* ... 45
21. *Dicarpa simplex* ... 46
22. Modifications in the branchial sac ... 49
23a–f. Modifications in the branchial sac ... 50
23g. Modifications in the branchial sac ... 51
24. *Ciona intestinalis* ... 57
25. *Clavelina picta*; *C. gigantea*; *C. oblonga* ... 58
26a. *Eudistoma olivaceum* ... 59
26b. *Eudistoma capsulatum* ... 60
26c. *Eudistoma capsulatum* ... 60
26d. *Eudistoma hepaticum* ... 61
26e. *Eudistoma carolinense* ... 61
26f. *Cystodytes dellechiajei* ... 61
27a–c. *Distaplia clavata*; *D. bermudensis*; *D. stylifera* 62
27d. *Distaplia clavata* ... 62
28a. *Aplidium pallidum* ... 63
28b–f. *Aplidium pallidum*; *A. glabrum*; *A. pellucidum*; *A. exile*; *A. stellatum* ... 64
29a–i. *Didemnum* sp.; *Trididemnum* sp.; *Diplosoma*; *Echinoclinum*; *Lissoclinum* ... 66
30a–e. *Perophora viridis*; *P. bermudensis*; *Ecteinascidia tortugensis*; *E. turbinata* ... 68
31a. *Rhodosoma wigleii* sp. nov. ... 69
31b. *Corella borealis* ... 70
31c. *Chelyosoma macleayanum* ... 70
31d. *Chelyosoma macleayanum* ... 71
32a–k. Species of *Ascidia* ... 72
33a–g. *Boltenia ovifera*; *B. echinata*; *Halocynthia pyriformis* ... 74
34a. *Pyura vittata* ... 76
34b. *Pyura vittata* ... 76
34c–e. *Microcosmus exasperatus*; *Pyura vittata*; *Cratostigma singulare* ... 77
35. Spread of *Cratostigma* ... 78
36. Species of *Molgula* ... 80
37a. *Molgula complanata* ... 81
37b. *Molgula complanata* tadpole ... 82
38. *Molgula occidentalis* ... 82
39. *Molgula siphonalis*; *M. griffithsii*; *M. retortiformis* ... 83
40. *Molgula arenata* ... 85
41. *Molgula arenata* ... 85
42a–f. *Polycarpa*; *Cnemidocarpa* ... 87
43a–g. *Botryllus*; *Botrylloides*; *Symplegma*; *Dicarpa*; *Polyandrocarpa* ... 88
44a–g. *Styela*; *Pelonaia*; *Dendrodoa* ... 92

45. Continental Drift 100
46a–c. Early development 104
47. Ascidian gastrula and larva 105
48. *Ascidia mentula* tadpole 105
49a–b. *Ciona* and *Halocynthia* tadpoles 105
50. *Molgula citrina* tadpole 105
51a. *Aplidium constellatum* tadpole 106
51b. *Perophora viridis* tadpole 106
52. Possible steps in the origin of the vertebrates 106
53. *Bdellostoma stouti* 107
54. Developing eggs of *Bdellostoma stouti* 107
55. Embryo of *Bdellostoma stouti* 108

COLOR PLATES

Frontispiece *Boltenia ovifera*
I. *Ciona intestinalis* 23
II. *Clavelina oblonga*; *Clavelina gigantea* .. 25
III. *Distaplia bermudensis* 26
IV. *Distaplia bermudensis* 27
V. *Aplidium (Amaroucium) constellatum* .. 28
VI. *Aplidium (Amaroucium) constellatum* .. 29
VII. *Didemnum candidum*; *Didemnum albidum* 30
VIII. *Aplidium pallidum*; *Trididemnum savignii; Diplosoma macdonaldi* 30
IX. *Boltenia echinata*; *Pyura vittata* 31
X. *Cratostigma singulare* 32
XI. *Molgula citrina*; *Molgula complanata* .. 33
XII. *Molgula arenata*; *Molgula manhattensis*; *Molgula retortiformis* 34
XIII. *Bostrichobranchus pilularis* 35
XIV. *Botryllus schlosseri* 36
XV. *Symplegma viride*; *Polyandrocarpa floridana*; *Polyandrocarpa maxima* ... 37
XVI. *Polycarpa obtecta*; *Styela plicata*; *Polycarpa circumarata* 38

TABLES

I. Key to the Orders and Families of *Ascidiacea* 16–18
II. Comparison of Branchial Sacs in Representatives of Ascidian Families 48
III. Compete List of Species of *Ascidiacea* Found 54–56
IV. Distribution of *Ascidiacea* in North America and Europe 95–96
V. Data Suggesting Ascidian Placement by Continental Drift 97

Preface

The phylum Chordata is universally used as the heading for the diverse and widely successful group of vertebrate animals, and it is easy to forget that two related subphyla are included which do not possess vertebrae. These are made up of widely distributed small marine species, the *Tunicata* and the *Cephalochordata*. The latter is a relatively small group containing a few species of the worm-like *Branchiostoma*, or Amphioxus, living in the sand in shallow water in all warm seas. These little species all possess a longitudinal axis rod, the notochord, and could easily be thought of as ancestors of fishes if rings of cartilage or bone (vertebrae) developed around the central axis. But certain other anatomical differences in pharynx and excretory systems in *Cephalochordata* suggest possible evolutionary regression from more advanced organisms.

In contrast, the *Tunicata* are more favored as vertebrate ancestors in the long past. They include three classes: the *Ascidiacea*, the attached "sea squirts"; the pelagic free-floating *Thaliacea*, including the jellyfish-like *Pyrosoma*, *Doliolum*, and *Salpa*; and the *Larvacea*, in which are species possessing in the adults a trunk and permanent tail like *Oikopleura*.

It is the sac-like sea squirts whose eggs develop into tiny, free-swimming tadpole larvae which seem to be closest to the anatomical organization from which the earliest vertebrates must have originated at the beginning of the Paleozoic era. Figures 1 and 2 show the structure and life history of the *Ascidiacea* in diagrams of a warm water Ascidian called *Ecteinascidia tortugensis*.

It is because of the imagined evolutionary position as probable vertebrate ancestors that the Ascidians have excited a wide interest for a hundred years. They are often bright colored, beautiful, but small animals growing on rocks or sand in shallow water or at moderate depths. They are found from just outside beaches to a hundred miles out to sea on the continental shelves of all continents from arctic to antarctic, and a few species are widely dispersed on deep sea bottoms in the basins of oceans all over the world. Since most are soft-bodied animals, there are few known fossils which have been found. Indeed, F. Monniot (1970) described a *Cystodytes* from Pliocene deposits of Breton. Only one fossil specimen has received questionable identification as an ancient Ascidian derived from Paleozoic strata. Although several hundred million years old, the specimen called *Ainiktozoon*, was described by Scourfield (1937) from Silurian of Cornwall. According to Hutchinson (1961), it appears like a pelagic Ascidian trying prevertebrate movement in the sea before the loss of the notochord. The actual evolutionary transition from Ascidian to primitive jawless vertebrate must have happened even earlier at the dawn of the Paleozoic epoch.

The study and identification of the many families and species of *Ascidiacea* began long before Linnaeus organized classification with his *Systema Naturae* in 1767. Indeed, Aristotle recognized sea squirts, and two that he described were given the same names by Linnaeus. These are *Ciona* and *Boltenia*, and just as from the North Sea in Linnaeus's time we still dredge them from the Gulf of Maine. Many important world-wide collections have been made, beginning with the long voyage of the sea exploration ship the *Challenger*, and its studies which resulted in the many volumes of *Challenger Reports* published from 1873 for forty years. Their reports on *Ascidiacea* were by Herdman, included in three volumes, 1882 and 1886. More restricted studies have reported and described the principal species from many locations on continental shelf sites all over the world. The kind of marine environments where the principal species live is ordinarily stated, and records are often given to show how extensively the species are spread. These facts about Ascidian species and geographic distribution have been summarized so that it has become possible to indicate how widely the more important species are spread throughout the seas of the world. The most notable of such attempts to show the distribution of commoner *Ascidiacea* species in the nothern hemisphere was made by Hartmeyer (1923) and covered the seas of the northern Atlantic, Europe, and North America. This has been extended for European seas by Ärnbäck–Christie–Linde (1934), by Huus (1937), by

Brien (1948), and by the many studies of Berrill (cf. 1950).

Additions to knowledge of the distribution of Ascidian species on the continental slopes and in deep seas have been made by several recent publications of Millar (cf. 1959) and of both Claude Monniot (1961) and Francoise Monniot (1965).

For the coastal locations of both North and South America, the works of W. G. Van Name (1945) are invaluable for the species descriptions and for the very complete bibliography for every species described. Many of his quoted records were made by specialists in other fields who contributed specimens and records to the collections of the American Museum of Natural History in New York, so the source data are not uniform. The 1945 volume summarized Van Name's lifelong studies of *Ascidiacea* at the Museum, and it is the most useful summary of North American specimens up to the present time. It gives outline sketches and identification descriptions for all the species described up to that date. With the more important species, Van Name gave notes of the bottom characteristics from which the species had been reported. Only for certain limited marine sites were the species distributions based on personal collections, and they were occasionally fragmentary or incomplete for some species. Studies which are based on firsthand examination of living specimens and which report distributions from methodical collections in planned areas spread along the whole length of the Atlantic continental shelf of the United States have not been made until the present study.

For this volume an attempt was made to collect and describe all the species of *Ascidiacea* found during the past five years from eight different representative bottoms on the continental shelf off the eastern United States. Five of these were checked more completely than the other three, but all involved dredging over areas at least a half mile square. In every case, dredgings were made, often many times. The locations sampled were as follows: I. Gulf of Maine. II. Nantucket Sound, Buzzards Bay, Naragansett Bay, Northern Long Island Sound, all with deeper water samples. III. Offshore sampling was done off New Jersey, Maryland, and Delaware. IV. Off-shore samples south of Cape Hatteras. V. Dredging off Georgia coast from shallow to deep water at the edge of the continental slope from South Carolina to Florida. VI. Off Tortugas Islands, west of Key West, Florida. VII. Florida west coast, Tampa Bay to Fort Myers. VIII. Off northern Florida, including Apalachee Bay. In addition, some preliminary surveys were made off Padre Island, Southern Texas, in the Gulf of Mexico. All these collecting areas, except the last, are indicated on the chart of the Atlantic continental shelf of the United States (Fig. 5).

The specimens were all identified by the writer, and in most areas the depths and bottom samples were examined at the collection points. In a very few cases, indicated when the particular species is discussed, specimens were classified when they were brought in by other collectors, but for these too the collection sites were verified with the collectors concerned.

Four new species or subspecies have been found in the course of these studies, and it is shown that the geographic distributions for several well-known species have been incompletely charted in the past. These facts are noted in the appropriate sections, but the detailed descriptions of the new species or subspecies will be given in a separate paper to be published later. This study of the *Ascidiacea* makes it possible to get a new start at the interpretation of the past history of the class on the Atlantic continental shelf of North America since the Mesozoic, and especially the influence on species distribution by the separation of Europe and North America by continental drift and the widening of the Atlantic ocean.

It is a pleasure to acknowledge the assistance of many different people and institutions in the work of completing this new summary of the "sea squirts" found on the continental shelf of the United States. There is a very real debt to Dr. Roland L. Wigley of the National Marine Fisheries Laboratory at Woods Hole, and the personnel of the vessel *Albatross IV*, whose initial collections furnished the stimulus for this new survey of the *Ascidiacea*. Additional material was collected by the Marine Biological Laboratory collecting personnel and the research vessel *A. E. Verrill*. Some assistance was received from Duke University Marine Laboratory at Beaufort, North Carolina, especially from Dr. C. G. Bookhout and Dr. John D. Costlow, and several specimens were made available for identification. There is a real debt to the Marine Laboratory of the University of Georgia at Sapelo Island. Their collecting vessel *Kit Jones* and the patience and understanding of Captain Rouse were both of great assistance. Special help was given by Scott Leiper of the Department of Zoology, who made available for study some of his specimens from in-shore dredging.

Two groups of marine investigators on the west coast of Florida deserve special thanks for their collecting of Ascidian specimens and their careful records of the dredge points. First there is the Marine Research Laboratory of the Florida Department of Natural Resources and Director William G. Lyons. He made available for study by the writer the Ascidian specimens collected in the Hourglass Cruises at a number of selected points off St. Petersburg and Fort Myer. Helpful assistance and a number of specimens are gratefully acknowledged from John Rudloe of Panacea, Florida.

Specimens collected at his Gulf Laboratory in Apalachee Bay, south of Panacea, were also gratefully received.

Finally, special thanks are due to Nancy Rabalais of Texas Agricultural and Industrial University, Kingsville, Texas, working as a graduate student on the specimens which she and her colleagues have collected from inside and outside of Padre Island, south of Corpus Christi, Texas. Several specimens not previously reported from the Gulf of Mexico were identified.

Another sort of assistance has come from permission to redraw or otherwise adapt a number of figures published in important Ascidian monographs, as well as some from the original descriptions. All named species have been examined, but it is sometimes more satisfactory to readapt figures already published by W. G. Van Name or N. J. Berrill than to draw new figures. So with permission of the publishers, the American Museum of Natural History, N.Y., several figures have been redrawn from W. G. Van Name's *North and South American Ascidians*. Three others were redrawn from N. J. Berrill (1929 and 1950). In the text an acknowledgment is made in the legend of each diagram in which a redrawn figure is included.

In addition to the sources of figures mentioned above, several figures have been redrawn from descriptive articles in several biological journals. Each of these is acknowledged in the legends to the figures. This study, just as every summary volume or taxonomic study, has received the conscious or unconscious assistance of systematic investigators who have previously worked on these or related species.

The author gratefully acknowledges the assistance for the completion of this new taxonomic study of the *Ascidiacea*, or "sea squirts," of a grant to Amherst College, Amherst, Massachusetts from the National Science Foundation (Grant No. GB6902X, 1969–72). The completion of the volume has been made possible by a grant from Charles Merrill Trust (1972–74) to the Marine Biological Laboratory, Woods Hole, Massachusetts. Each of these grants has kept me active in the survey and in the publication of the report. I regret that unavoidable delays have held up the final publication of the volume. The patience of the officials in the administering institutions has made a real contribution to the final publication of the work.

One final word of appreciation is due to the artist, Mareile Fenner, for her patience and long-continued work in completing the family plates and the line sketches of every one of the species of *Ascidiacea* which has been seen. She has drawn the specimens either whole, dissected, or mounted for microscopic study. Although not familiar with any of the species before the study began, as time went on she often noted anatomical details which the author had missed. Her patience and skill have furnished the most distinctive and useful part of this descriptive volume, setting forth again the "sea squirts" of the Atlantic continental shelf.

I. The Ascidian Structural Plan

There are many reasons for the widespread interest of zoologists in the *Ascidiacea* or "sea squirts." They are exclusively marine and they are found as benthic or bottom-living animals in all seas from subarctic to the tropics. Some species thrive living attached loosely to sand in the cold boreal current in Davis Strait between Greenland and Labrador. Others survive best lying loosely on sand in the warm Gulf Stream current, rich in tiny food particles, fanning northeastward from the Straits of Florida between the peninsula and the island of Cuba.

They are unique animals attached by their sac-like tunic which contains tunicin, a substance chemically almost identical with cellulose. Through the tunic the mantle-covered body maintains contact with the sea water by means of two siphon openings, incurrent or oral, and excurrent or atrial. Water is drawn through the incurrent opening into a large branchial sac pierced by rows of slits, or stigmata, which are lined with beating cilia. The water passes through these stigmata into the atrial cavity surrounding the branchial sac and out from the excurrent siphon. Tiny food particles are collected in a mucus roll inside the branchial sac and passed into the esophagus (Fig. 1).

The gut tube continues into the stomach and on into the intestine, which loops back to open into the water of the dorsal atrial cavity. Gonads are usually placed in the intestinal loop, and from these sperm and oviducts pass into the atrial cavity. Excretion is carried partly by cells of the stomach wall and partly by short tubules attached to the intestines. Other cells appear to concentrate urates within small tubules, or among the *Molgulidae* in a larger reservoir, the kidney. It appears that all tissues excrete ammonia into the sea water. Each living cell carries on most of its essential physiological activities independently of the rest.

There is an open blood system. A tubular heart lies below the digestive loop and usually bends on itself to become a U-shaped tube. Blood from the digestive tract is collected in the ventral sinus below the branchial sac and forced up through vessels between the stigmata. After several minutes the heartbeat slows and stops. Following a pause it starts beating again in the opposite direction, from the dorsal side downward into spaces around the viscera. This reversal is a unique physiological activity in the heart of the *Tunicata.*

The blood system carries some digested food material, some oxygen, and a store of inorganic ions. As an open system it is less efficient than the blood system of the vertebrates. It carries on some unique transformations which seem physiologically experimental, the advantages of which are not certainly known. One such activity of the blood systems of some rather primitive

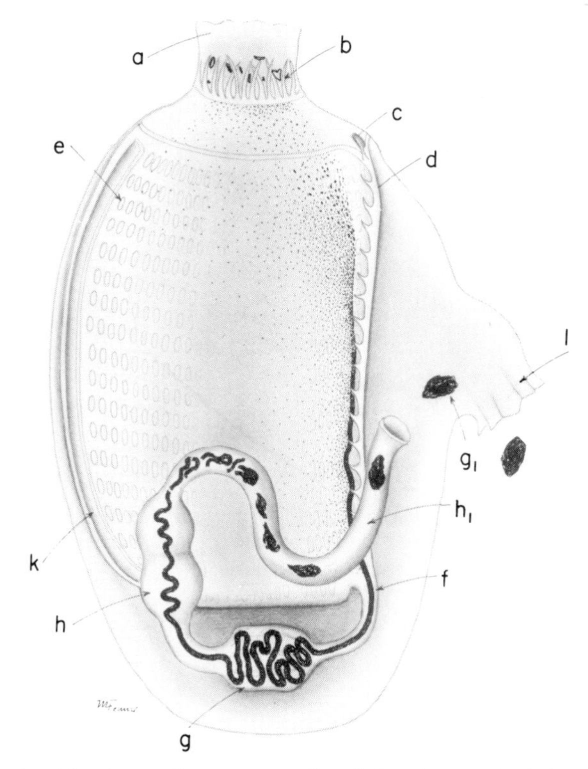

Fig. 1. *Ecteinascidia tortugensis* adult. (a) oral siphon, (b) oral tentacle, (c) dorsal tubercle, (d) dorsal lamina, (e) stigmata, (f) esophagus, (g) stomach, (g_1) excreted particle, (h) intestine, (h_1) rectum, (k) endostyle, (l) atrial siphon.

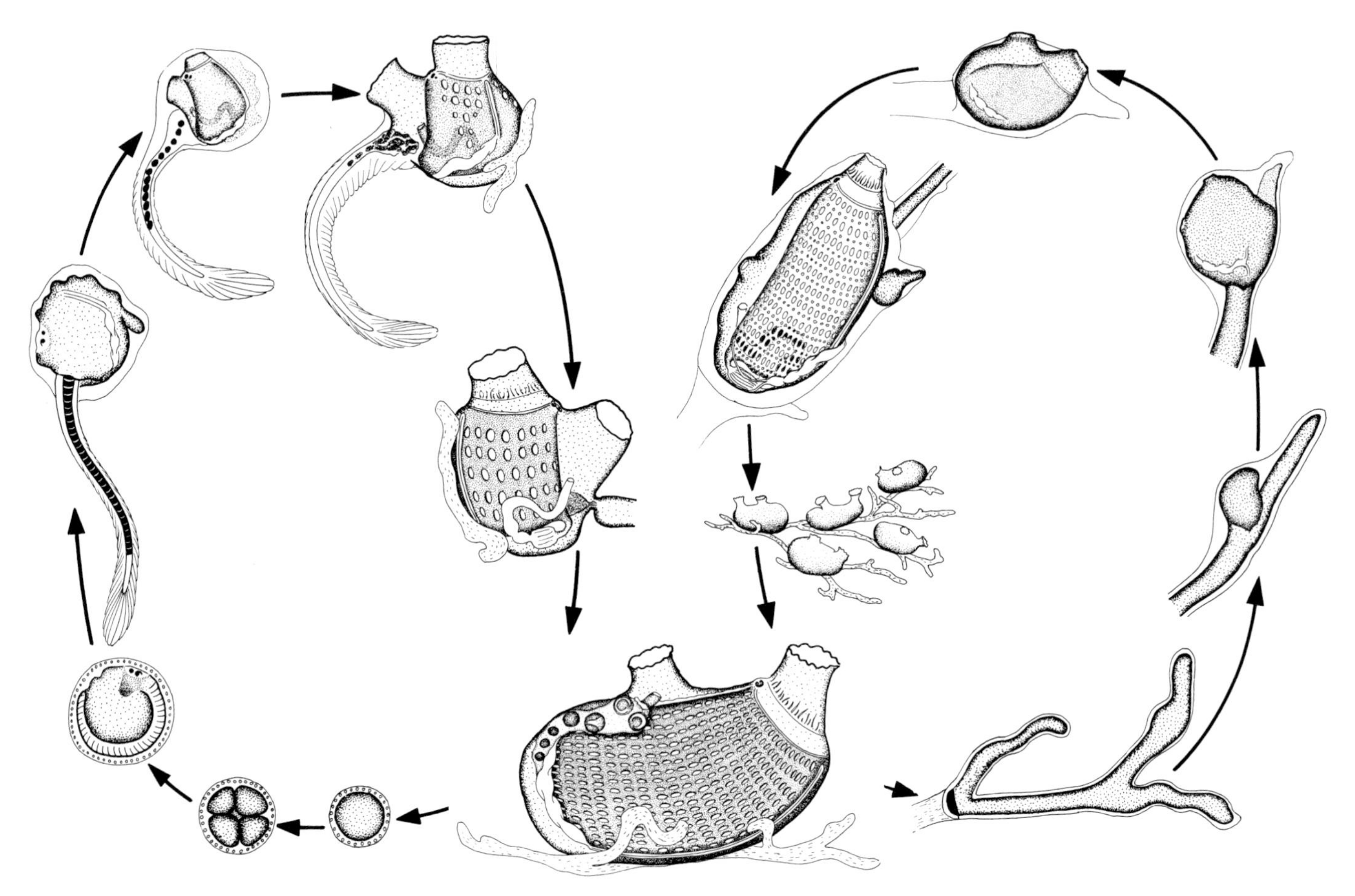

Fig. 2. *Ecteinascidia tortugensis.* (Left) Development of egg to tadpole, and metamorphosis to adult. (Right) Vegetative cycle, with budding to form a colony.

Ascidians is the great concentration of vanadium from sea water (cf. Goodbody 1957, Barrington 1965). This seems to have no respiratory activity. In other more advanced species there comes about concentration of certain other metals such as iron or niobium, but the significance of these physiological experiments with metals in *Ascidiacea* has not been established.

The nervous system consists of a cerebral ganglion located inside the mantle between the two siphons. This is connected with a sublateral gland usually below and by nerve fibers to the oral siphon and other muscular tissues. The sublateral gland opens into the entrance to the branchial sac by a funnel-shaped dorsal tubercle, and it influences the siphon currents.

The food-gathering mechanism of the *Ascidiacea* also is unique. They are all ciliary or suspension feeders. The groove on the ventral side of the pharynx, the endostyle, secretes mucus into the cavity whenever water is being drawn in. This mass is continuously revolved by ciliary action within the branchial sac and collects all the fine particles of food in the sea water in the sac into the mucus. The mucus mass is drawn upward in the branchial sac and payed like a rope into the esophagus opening at its posterior end.

The drawings of *Ecteinascidia tortugensis* (Figs. 1, & 2) show all the anatomical features mentioned above. Compare also Figure 24 of *Ciona intestinalis*, which shows the same anatomical arrangement in a young more northern species.

Eggs are fertilized at the upper end of the oviduct and develop in the atrial cavity to the tadpole stage. In many other species like *Ciona* the eggs pass directly into the sea water and develop there from the start. The tadpole larvae of *Ecteinascidia* (Fig. 2, left) are distinctive, with two head sense organs—a vibration-sensitive otolith and an eye spot or ocellus—four rows of stigmata and a tail with a fine dorsal nerve cord, and the thicker axis rod or notochord. The larvae swim a few minutes to an hour or so, then become attached at the front end. The tail is then rapidly resorbed, and the two siphons break through. The mouth of the larva does not break through and so it does not feed in any species.

Vegetative budding goes on in *Ecteinascidia* at the same time that eggs and larvae are developing (see Fig. 2, right). The ventral vessel covered by ectoderm forms a growing tip under the zooid, which continues to grow out as a long branching stolon. At intervals thickenings appear which enlarge as buds. From each a new zooid develops with two siphons like the parent. In *Ecteinascidia* and *Ciona* all zooids remain interconnected by stolons in a branching colony with the blood current flowing through the stolons from one zooid to another. In other simple Ascidians the zooids separate from the parent and remain independent individuals. Finally, many simple Ascidian species never bud at all, but remain single zooids as formed from the egg.

II. Distribution of the Species of *Ascidiacea* in the Gulf of Maine

The widespread distribution of Ascidians as benthic animals in seas all over the world has suggested to marine zoologists that this is an ancient group of animals, even though fossil records are nearly nonexistent. Ascidian development with the widespread appearance of a tadpole larva was described in 1847 by P. J. van Beneden, but they were universally classified with the *Mollusca*. It was the recognition by the Russian zoologist Kowalewsky in 1867 of the typical chordate character of the larva (reproduced in Fig. 48), which stimulated the interest of biologists and resulted in many studies of Ascidian species in all parts of the world. These have included such extensive surveys of the seas of the world as the *Challenger Reports* in England (Herdman 1886), the studies of A. E. Verrill on species of the American northeast (1870–74), the examinations of many museum specimens by Ärnbäck–Christie–Linde (1924), the developmental sequences by Brien (1948), Van Name's wide sampling of species on the North American littoral, Berrill's many studies of tunicate structure from 1929, and the description of deep sea material from the Danish *Galathea* voyages (Millar 1959). It has become widely recognized that these apparently insignificant, tadpole-forming, benthic animals have had a long and ancient history (cf. Meglitsch 1967).

Tunicata include three classes of marine animals: *Ascidiacea*, *Thaliacea*, and *Larvacea*. The second of these includes the pelagic *Salpa* and *Doliolum*, and the third the world-wide floating species of Appendicularians. The last two are highly modified species, the recent relationships of which are difficult to determine. The similarity in body plan of all three classes makes it reasonably certain that they all were derived from similar basic protochordate stocks in the long past (cf. Berrill 1955 & Lohmann 1933). The *Ascidiacea*, with their swimming chordate larvae and fixed adult stages, show several directions of evolutionary change which it is profitable to follow briefly before details of species differentiation are studied. Species of these bottom-dwelling sea squirts are found crowded in many tidal sandy or stony sites close to shore. Other species on the bottom are loosely attached to shell or stone by tiny filaments formed by the tests. Still others in water of twenty to fifty meters have rigid attachment by stalks to stones, shells, or underwater cliffs. Still others hold a precarious attachment on the continental slope. Finally, a few tiny modified species are distributed in dark but food-rich water on deep sea bottoms a thousand or more meters down.

In spite of the apparent similarities in their suspension feeding apparatus, Ascidian species maintain a constant interaction with their immediate environment. Species such as these which live in sea water currents rich in food particles possess branchial sacs. All are influenced by the temperature range. All are dependent on ocean currents, and several have developed hold-fast organs on the tunic. Some groups of species living in much the same sorts of bottom environments have developed a mutual interdependence with other attached Ascidian species, or with other marine animals in the area. Species living in tidal areas may be so numerous as to form thick masses of individuals and so a surer hold-fast for all. Other Ascidian species have taken up a settled attachment on sand flats at moderate depths many miles at sea on the Atlantic continental shelf. So such species as *Molgula arenata* may be much more numerous than any of the better known Ascidian species close to shore.

It has been demonstrated that different species of *Ascidiacea* occupy just as well-defined areas on the sea bottom, conditioned by surface conditions, ocean currents, and land and water temperatures as do insect species on land. Occasionally two unrelated species are intermingled on the same bottom area, but more often they have slightly different preferences, even when close together in similar habitats. Both *Cnemidocarpa mollis* and *Styela partita* can be dredged from a half mile off shore at Chatham, Massachusetts, but usually the former will be in deeper water, even out to Georges Bank, and the latter can be found on piling of the on-shore wharf.

Boltenia ovifera is a moderately deep water species attached by its long stalk to rock or shell in the cool currents of the Gulf of Maine. It is a common species on the outer part of Georges Bank and also closer to shore off Mt. Desert Island and Nantucket Shoals. Here also is found the chalky encrusting colonial species *Didemnum albidum* widely attached to stones and shells. In mud pockets in deeper water, bathed by the cold current making in south of Nova Scotia, are specimens of *Polycarpa fibrosa.* Farther off shore at medium depths in the Gulf of Maine are found the tough *Ascidia prunum* and the related *Ascidia callosa.* Finally, in shallow water at a half mile off shore all the way from the Bay of Fundy to Cape Cod Bay, another widespread northern species, *Molgula siphonalis*, may be picked up on the sand. Distribution plots for a number of these species are shown in Figure 3.

In contrast, are the distribution plots of a number of well-known species along the whole Atlantic continental shelf later, in Figure 5. Note especially *Aplidium constellatum*, *Molgula manhattensis*, *Styela partita*, and *Bostrichobranchus pilularis.* Each of these species shows a wide distribution which can be contrasted with those of much more limited range. Compare with those cited above the many species distributions in limited areas like *Molgula occidentalis* on the chart of the whole Atlantic shelf later, in Figure 10.

Even though the branchial sac structure and the food and oxygen intake are mechanically similar in all Ascidian species, it is apparent that small differences in the food brought and the oxygen carried by the water over different portions of the sea bottom make some ranges very favorable for certain species and unacceptable for others. In addition, collections of Ascidian specimens of common species made within certain well-marked sections of the Northeast from successive collections made in different decades show certain differences in species distributions. These differences suggest that slow migrations may have occurred and probably are still occurring. The most frequent differences are found in the records of the northern boreal interconnecting zones between Europe and North America, and they give evidence of a past continental movement rather than a recent species transfer.

Uncertainty about migration of particular species is more frequently caused by errors in species identification than by evidence of actual migration of species in time. Collections of Ascidians and many other marine animals were made by the industrious veteran collector A. E. Verrill off New England and Canadian coasts from 1870 to 1875, and intermittently later. A tabulation of Ascidians was given in "Catalogue of the Marine Invertebrates of Eastern Canada" in 1900 by Whiteaves. In addition, Sumner, Osborn, and Cole (1913), under Bureau of Fisheries auspices, published "Biological survey of the waters of Woods Hole and vicinity," and this was rechecked by W. C. Allee and his associates of the Marine Biological Laboratory in 1923. Meanwhile a major revision of Ascidian classification had taken place leading eventually to acceptance of the orders and suborders of Garstang (1928) which we now use. At about that time Berrill (1928) began his long and influential studies dealing with Ascidian species and their larvae. Following these and many other systematic studies Ascidian species classification for the American shelf species was reasonably well stabilized with the definitive monograph of W. G. Van Name in 1945, *North and South American Ascidians.* Although this is too voluminous for use by beginners in Ascidian species identification, it remains the most useful and trustworthy catalogue of the species found on the Atlantic continental shelf of North America and the West Indies. It is from this point that the questions brought up tentatively by Allee, by Berrill, and recently by C. Monniot (1969) and others discussed in the present study need to be reexamined.

A careful tabulation of the records showing the species distribution of the more important Ascidian species of the northern hemisphere as reported in more recent publications was made by Hartmeyer (1923) in his volumes describing the records of the collections of the Danish Ingolf Expedition. It is especially fortunate that there are now becoming available for comparison with the Hartmeyer records a whole series of dredging locations for Ascidian specimens such as those found in the Gulf of Maine during the past five years by the Benthic Investigation staff of the laboratory of the U.S. National Marine Fisheries at Woods Hole, Massachusetts. The methods used for collecting and the ocean areas covered in the Gulf of Maine may be summarized for comparison with earlier records. The survey cruise made by the research vessel *Albatross IV* in the summer of 1968 was one of the most useful of the collecting cruises, not only for tracing the fish but because of the information given or confirmed about smaller animal groups like the *Ascidiacea.* It enabled the gathering of new ideas on Ascidian species distribution in the Gulf of Maine and rechecked previously recorded observations on well-distributed species. For instance, it was amply reconfirmed that *Polycarpa fibrosa* is a tough, deep water species living in mud pockets south of Nova Scotia from which secure attachment points its elastic siphons are extended to get food particles more than five times the length of the animal's body. These extensive *Albatross IV* Ascidian species records can be compared with those from much earlier dredgings to indicate distribution changes in fifty or a hundred years.

The *Albatross IV* Benthos Collecting Cruise of 1968 (Alb. 68–12) was conducted under the scientific direction of Dr. Roland L. Wigley. It left Woods Hole, Mas-

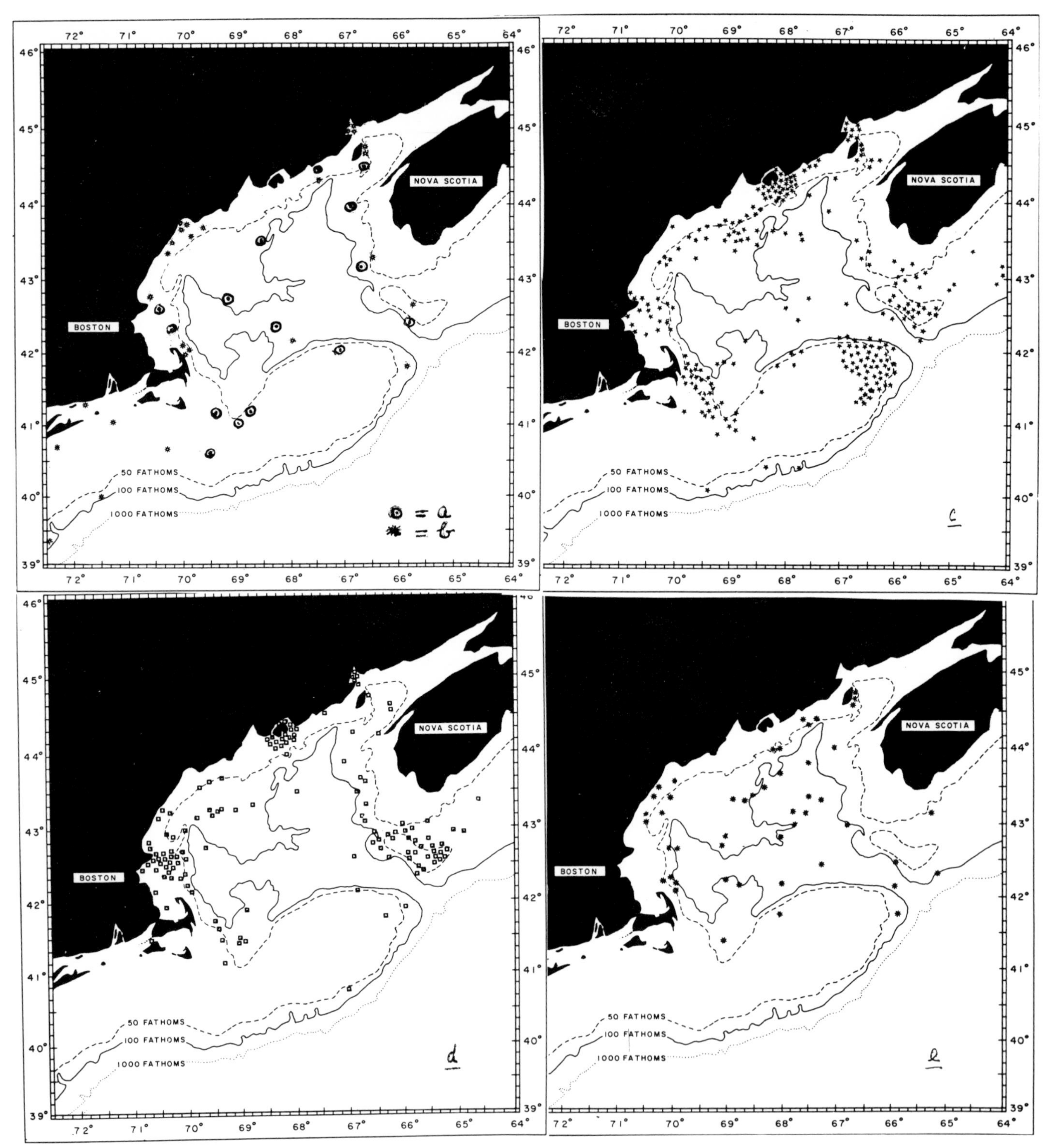

Fig. 3. Distribution of five common species found in the Gulf of Maine. (a) *Didemnum albidum*, (b) *Molgula siphonalis*, (c) *Boltenia ovifera*, (d) *Ascidia callosa*, (e) *Polycarpa fibrosa*.

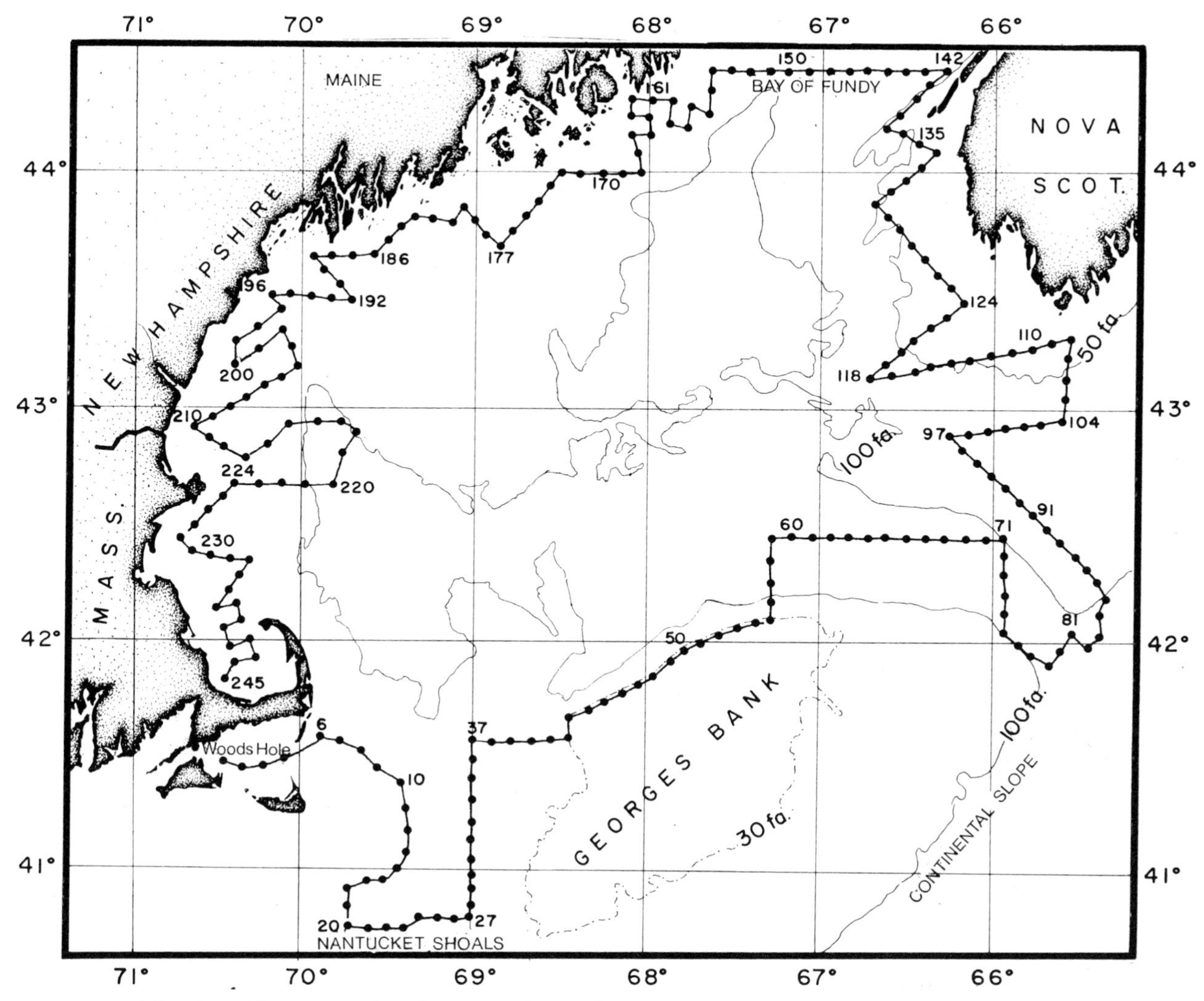

Fig. 4. *Albatross IV* benthos-collecting cruise of 1968. The numbered round dots were dredging points.

sachusetts, on August 12, 1968, and returned on August 22. The course is shown plotted on a U.S. Coast and Geodetic Chart in Figure 4. The weather was exceptionally favorable and the ship was in motion from the start until it returned to port. Every round dot on the plotted course was the site from which one or more dredged samples was brought from the bottom. The total number of dredgings was 245 and nearly all of them yielded specimens for study. Naturalist dredges of different widths and weights were used throughout and all were lined with quarter inch mesh nets. The shallowest dredging was from 20 meters depth at Crab Ledge, north of Pollock Rip just east of Cape Cod. The deepest was from about 1420 meters down east of the tip of Georges Bank at the entrance to the Gulf of Maine. This last point was off the shelf and on the Maine continental slope.

The dredge load from the bottom was dumped on the after deck of the *Albatross*, where it was spread with shovels and sorted over by the scientific teams on duty. The frontispiece shows some of the stalked Ascidians, *Boltenia ovifera*, collected from Station 177, east of Penobscot Bay, as they appeared alive and in color. The contents of dredge #177 was drawn up from the bottom at 60 meters and emptied on the deck for sorting.

The scientific personnel was divided into watches so that some were on duty at all times of the day or night. The material was roughly sorted out into major animal or plant groups. Some were held alive for study and others were preserved in formaldehyde. Most specimens were transferred to alcohol a day or two later. In addition, large samples of sand or mud and some of the matted material from the bottom were occasionally preserved for study. The preserved material was brought to the laboratory of the National Marine Fisheries Service and continues to be studied by specialists in some of the major groups for several years after collection.

The *Albatross IV* 1968 trip has proved to be a very profitable collecting expedition for accurate plotting of the distribution of Ascidian species found on the bottom in the Gulf of Maine. The study has been continued with specimens brought in by collectors from the Supply Department of the Marine Biological Laboratory, and on material collected by the writer on trips of the research vessel *A. E. Verrill* of the Systematics and Ecology Program, now largely discontinued. The extension of the study of species of *Ascidiacea* found on the Atlantic continental shelf south of the Gulf of Maine is described in later chapters.

The "sea squirts" or Ascidian specimens are basically similar in structure whether they are tiny semitransparent one millimeter encrusting zooids in colonies, or giant single zooids five inches in length. A *Molgula* from the bottom on Georges Bank in the Gulf of Maine is similar in structural plan to a *Molgula* found on the shelf off Georgia. With the basic plan of Ascidian structure of any species in mind, one can study its distribution in any area of the United States shelf.

III. Biogeographical Regions on the Continental Shelf and Distribution of Ascidian Species

A chart showing most of the Atlantic continental shelf of the United States is drawn in Figure 5. It includes the long and varied Atlantic coastline of more than a thousand miles, including the Atlantic Ocean and the Gulf of Mexico. It covers three major regions which influence the climate and, indirectly, the kind and distribution of the Ascidian fauna.

The *American Atlantic Boreal Region* includes the sea areas from Labrador south to the Gulf of Maine and east of Cape Cod. Many species are common to the European Boreal, and some to the arctic and the coastal areas of Greenland.

The Atlantic continental shelf south of Cape Cod to the Florida peninsula is the *American Atlantic Temperate Region*. This is designated as the *Virginian Province*, from Cape Cod to Cape Hatteras, and the *Carolinian Province*, south to Florida from that important division point. A number of Ascidian species found off the west coast of Florida, and in the western Gulf of Mexico are common inhabitants of the *West Indian Region*. These regions are apparent in the figure, where the major collecting sectors for this study also are marked.

Most species of *Ascidiacea* release eggs which develop into free-swimming tadpole larvae. These are ordinarily minute and swim for a short distance only, perhaps one hundred times the length of the parent zooid. The evidence indicates that swimming tadpoles do not greatly increase the distribution range of Ascidian species, any more than do the eggs floating in the sea water. It is obvious that the oversize tadpoles found in *Distaplia* or *Didemnum* which usually remain inside the tunic wall where the tail vibrates, do not extend the range of the adult species at all. The tadpoles are much more useful to the species as seekers of favorable sites like those of the parent zooids for attachment of the newly developing eggs. Temperature, depth, bottom characters and marine currents are all important in locating Ascidian zooids where they will have spots similar to their parents for collecting minute food particles from the water. The finding of locations for attachment which resemble those of other attached members of the same species appears to be the major factor determining a desirable bottom area for these small benthic "suspension feeders."

In Chapter II, a chart was given showing the collecting cruise of the research vessel *Albatross* in 1968 (Fig. 4). The dredging records of this and other fisheries vessels have been combined to furnish species distribution charts for five of the best known Ascidian species which were common in the Gulf of Maine during the past five years. These are shown in Figure 3. Each of these is an example of a species found widely distributed in the American Atlantic Boreal Region, and their range does not extend very far south of Cape Cod. There are other species found in the Gulf of Maine, but also living much farther south on the Atlantic continental shelf. Some of these latter species are shown in Figure 5, Location I, the Gulf of Maine.

By similar methods the distribution of many other Ascidian species have been recorded and plotted for seven additional major sections of the American shelf in the American Atlantic Temperate Region. These major collecting areas along the whole Atlantic continental shelf have been roughly combined in the one chart, Figure 5, which summarizes the distribution of twenty common Ascidian species. In addition, a few deep water specimens from the continental slope were collected by the Woods Hole Oceanographic Institution vessel, the *Gosnold*, and turned over to the National Marine Fisheries for their collection. These and other *Gosnold* collections have been especially valuable to supplement the collecting from sites east of New York, New Jersey, and Maryland (Locations II & III).

The three well-studied areas on the southern portion of the Atlantic shelf for this Ascidian survey are: the shelf off Georgia, especially east of Sapelo Island (Location V); the shelf off the west coast of Florida, especially between Tampa Bay and Fort Myers (Location VII); and the broad shallow water over the shelf south of Panacea, Florida, called Apalachee Bay (Location VIII).

That part of the continental shelf in the region south of Cape Hatteras in which studies of Ascidian distribution have been made similar to those in the Gulf of

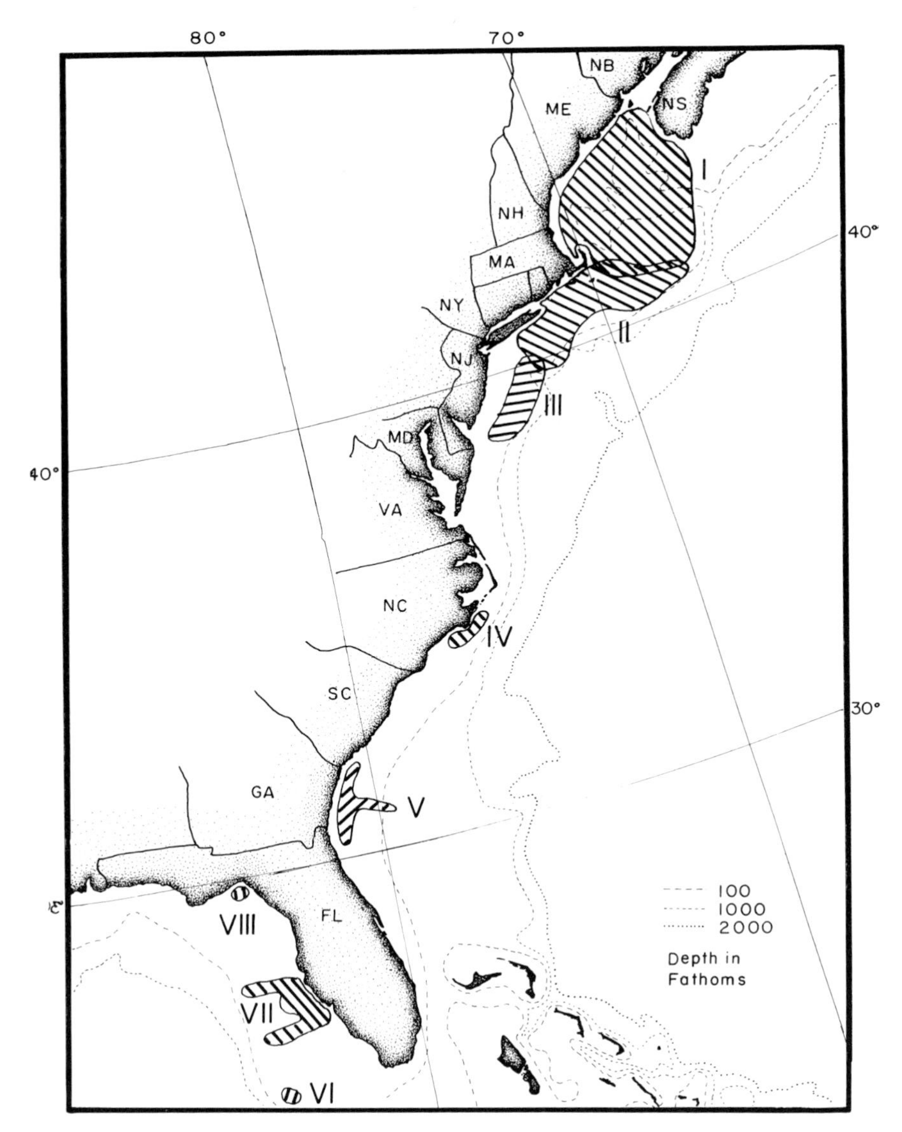

Fig. 5. Collection regions off the eastern United States (I–VIII) and twenty species (1–20) found there.

I. Gulf of Maine: 1, 5, 9, 10, 12, 13, 15, 16, 17, 19, 20.
II. Nantucket to Long Island: 1, 5, 7, 9, 10, 13, 15, 17, 19, 20.
III, New Jersey, Delaware, and Maryland: 1, 5, 7, 13, 15, 17, 20.
IV. Beaufort, North Carolina: 5, 7, 11, 13, 15, 17, 19, 20.
V. Sapelo Island, Georgia: 2, 4, 5, 6, 7, 11, 13, 14, 15, 17, 18, 20.
VI. Tortugas Islands, Florida: 2, 3, 4, 6, 8, 14, 18.
VII. Florida West Coast: 2, 3, 4, 5, 6, 7, 8, 11, 14, 15, 17, 18, 20.
VIII. Apalachee Bay, Florida: 2, 3, 4, 5, 6, 7, 8, 11, 13, 14, 17, 18.
Species: (1) *Ciona intestinalis*, (2) *Clavelina picta*, (3) *Eudistoma olivaceum*, (4) *Distaplia bermudensis*, (5) *Aplidium constellatum*, (6) *A. exile*, (7) *Didemnum candidum*, (8) *Ecteinascidia turbinata*, (9) *Chelyosoma macleayanum*, (10) *Ascidia prunum*, (11) *A. nigra*, (12) *Halocynthia pyriformis*, (13) *Molgula manhattensis*, (14) *M. occidentalis*, (15) *M. arenata*, (16) *Dendrodoa carnea*, (17) *Styela partita*, (18) *S. plicata*, (19) *Cnemidocarpa mollis*, (20) *Bostrichobranchus pilularis*.

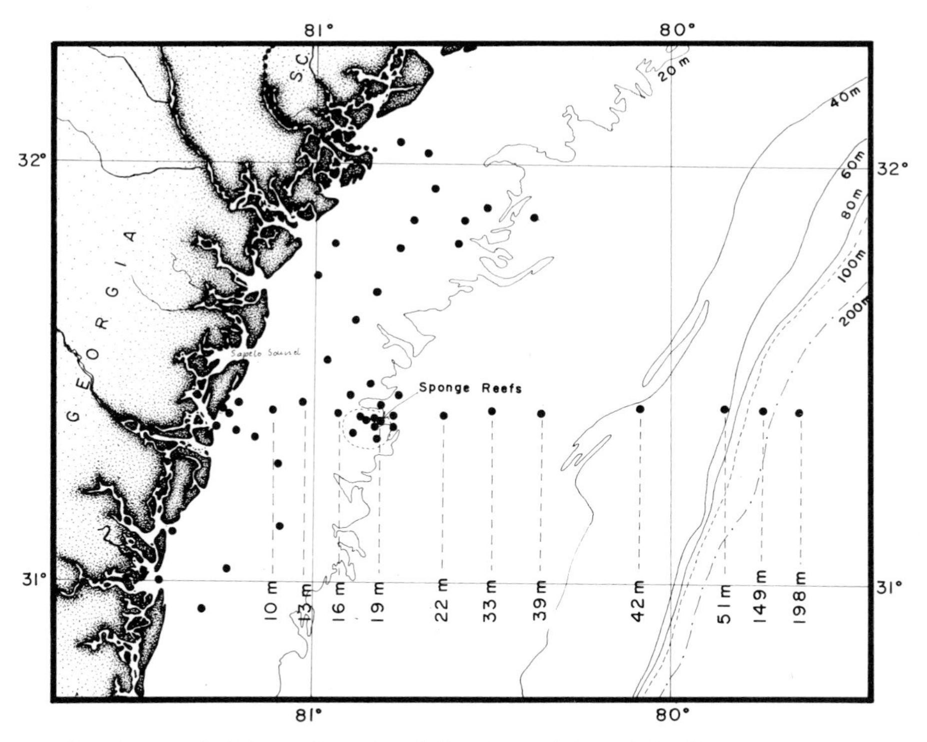

Fig. 6. Sapelo Island, Georgia. Collection points and depths.

Maine is the Georgia coastal region. This includes that portion of the shelf east of the Georgia coast, between the latitudes 31°–32° N and longitudes 81°–82° W, north and south of Sapelo Island. On Sapelo the University of Georgia has developed a Marine Laboratory which is served by the research vessel, the *Kit Jones*. In this vessel, with the helpful assistance of Captain Rouse and other members of the staff, it has been possible to collect Ascidian specimens from Sapelo Sound northward to north of Savannah and south to Jacksonville. In addition, the width of the continental shelf has been dredged from just north of Sapelo at one half mile off shore due east to the edge of the continental shelf at one hundred miles away. At that point the bottom depth was well below the one hundred fathom line. The Gulf Stream fans out northeastward over this shelf.

Figure 6 shows the coast and the continental shelf off Georgia. On it are plotted all of the dredge points where specimens were collected, from close in-shore to the edge of the continental shelf a hundred miles eastward. The record includes all the Ascidian species which I have found and identified and, in addition, several specimens collected by other zoologists and given to me for identification. As the list of species beneath the figure shows, there were sixteen Ascidian species collected from the Sapelo Island marine area. The circle of points labeled "Sponge Reefs" shows the location of a center at which many species are aggregated throughout the year. From that aggregation center eastward, the commonest species collected was *Molgula arenata*.

Several specimens and records methodically noted by Milton Gray at the Sapelo Laboratory I have identified. It is a pleasure to acknowledge Mr. Gray's contribution to an accurate catalogue not only of the *Ascidiacea* but to all the other phyla collected at Sapelo Island Laboratory. He was as careful and precise in recording the situation and time of the collection for the *Ascidiacea*, which he did not know well, as he was for the many groups of marine animals with which he was very familiar. It is important to record his contributions to forming an accurate catalogue of all marine species to be found off the Sapelo Island Laboratory.

Scott Leiper of the University of Georgia made systematic collections about one-half mile off shore at Sapelo Island beginning in the year 1971. In connection with his survey, he located and preserved a number of small specimens attached to shells and stones. Among these were some tiny interstitial Ascidian specimens which the writer identified as *Pyura vittata*. Drawings of these unexpected and interesting specimens are shown later in Plate IX and in Figures 34a, b, & d_{1-2}. In addition to these tiny specimens, the author has found this species several times larger and all the way to adult size of about 25 mm about one mile off shore in water about 10 meters deep. Leiper's studies show that *Pyura vittata* is found as immature zooids from close to shore to one mile out, where there is hard sand. It is found also in somewhat deeper water from the Tortugas Islands north along the west coast of Florida.

Several additional collecting locations were made available in the course of this Ascidian survey. A number of Ascidian species brought into the Duke University Marine Laboratory at Beaufort, North Carolina (Location IV), were identified. More than twenty-five years ago this investigator (Plough & Jones 1939) collected and published a list of Ascidians found off the Tortugas Islands west of Key West (Location VI). Since the Carnegie Marine Laboratory is now closed, it is necessary to arrange a special boat to get to the Tortugas at the present time. Even a brief visit showed much the same assemblage of Ascidian species seen thirty years ago.

In 1970 an important series of collections in all the phyla of marine animals was made by the staff of the Marine Research Laboratory, Florida Department of Natural Resources at St. Petersburg, Florida (William G. Lyons, director). The study was designated as the *Hourglass Cruises* and the essential data are available in the first number of the reports by E. A. Joyce, Jr., and Jean Williams (1969). They established twenty fixed off-shore stations off the Florida west coast in 1966 and 1967. Monthly collecting trips were made from ten stations due west of Tampa Bay and Fort Myers (Sanibel Island) on a six-day schedule. They used a variety of collecting gear, including a bottom sampling dredge. Typical samples of Ascidian specimens at each station were sent to me for identification. These came from depths of 6 meters to 75 meters on the continental shelf.

In 1971 I visited the laboratory at St. Petersburg, and reexamined the Ascidian collections from each of the noted localities. This turned out to be an unusual opportunity to study new and methodically collected Ascidian specimens from the Florida west coast. This study gave me additional material collected in an area not recently checked for *Ascidiacea*. For this I am greatly indebted, and the study has been extended by the kindness of the Director. The providing of specimens collected at a series of carefully measured depths at different seasons of the year, and in a geographic situation which has not been checked for the distribution of Ascidian specimens, was a valuable supplement to the present Ascidian study. Probably the most valuable data were records of collections at a series of carefully measured depths at different seasons of the year. The collection points of the *Hourglass* study off the Florida west coast are shown in Figure 7. They afford additional distributional data for the comparison of Ascidian species over wider marine areas off the west coast of Florida than has been available up to this time. In addition, there is clear evidence that each species tends to favor a particular depth.

Finally, yet another marine area has been examined for species of Ascidians on my visit to Panacea, Florida, with the cooperation of John Rudloe of the Marine Laboratory. He and his crew scraped the bottom at a succession of sites in the shallow waters of Apalachee Bay at the north end of the Florida west coast and brought in a series of distinctive shallow water specimens for my study in the laboratory. The locations and species collected are shown in Figure 8b. On the opposite side of the Gulf of Mexico, a few Ascidian species were collected at Padre Island, Texas, from shallow water inside and outside of the island (Figure 9).

To summarize then, the different regions on the Atlantic continental shelf from which collections of Ascidian specimens have been made are shown in a broad view in Figure 5. Successive portions of this long continental shelf are shown in more detail in Figures 3, 6, 7, 8, and 9. It should be understood that the species from these different regions often show differences in the assemblages at corresponding depths from one area to another.

Thus, in the Gulf of Maine, as noted in Figure 3, the species most commonly dredged are *Boltenia ovifera*, *Ascidia callosa*, and *Polycarpa fibrosa*. But off the Georgia coast, shown in Figure 6, species such as *Molgula arenata* and *Cnemidocarpa mollis* are those most likely to be found. It is easy to see the marked differences in the character of the two areas. First, the two sections are ten degrees of latitude apart. The water temperatures show an average difference of ten degrees in summer and often twenty degrees in winter. The same storms pass over each area, but they are usually more severe in the northern region. The Gulf of Maine is a great pocket protected by the bulk of Nova Scotia. The Georgia coast is only partly protected by the Carolinas on the north.

In addition to temperature averages, the two regions differ in bottoms and in marine currents. The bottom of the funnel of the Gulf of Maine is broad, but dissected by cold, deep water entering south of Nova Scotia and passing inward as far as a few miles off Cape Ann. Even more important is that the prevailing ocean current is from the north. It floods in counterclockwise direction into the Gulf of Maine, and bends in almost a circle to pass outward to the east at about the level of Cape Cod. The fisherman's favorite location, Georges Bank, lies roughly at the south border of the counterclockwise arctic current. Obviously, the effect of this is to make the water in the Gulf of Maine cold. It is fairly well loaded with microscopic food particles on which attached Ascidian species can well subsist, but the species are cold-water forms all the year.

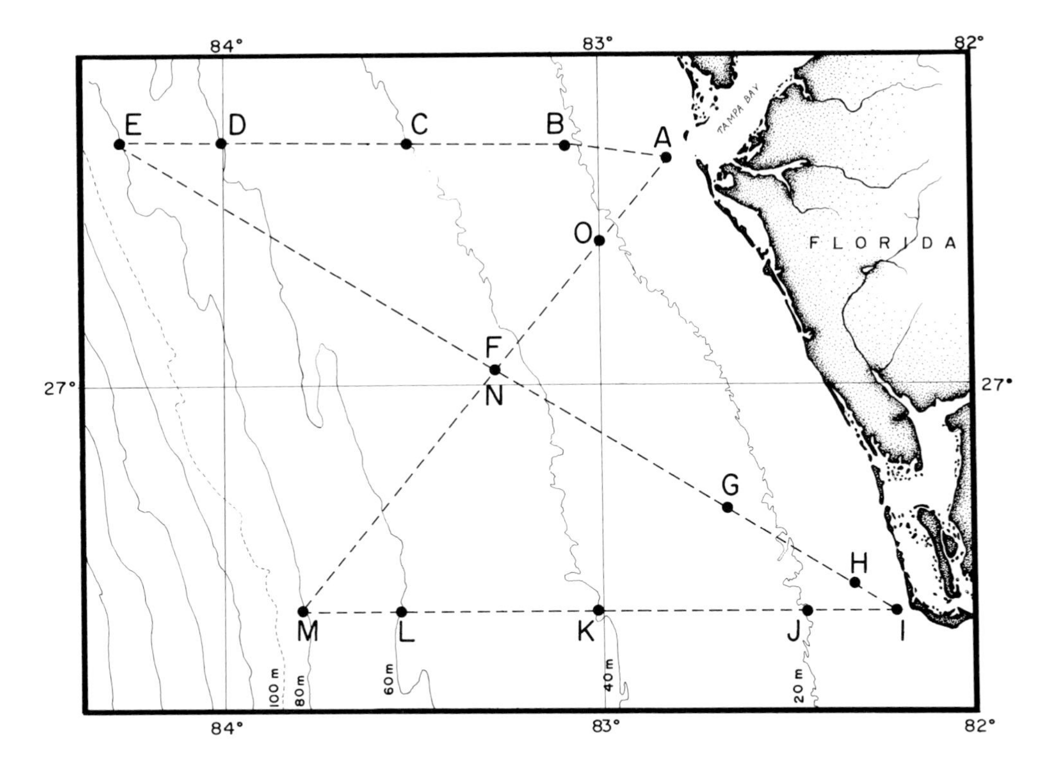

Fig. 7. Florida west coast from Tampa Bay to Sanibel Island. Collection points, depths, and species from the *Hourglass* cruises, 1965–67.

A. *Distaplia bermudensis, Clavelina oblonga, Molgula occidentalis, Styela plicata.*
B. *Styela plicata.*
C. *Polycarpa obtecta, Symplegma viride, Trididemnum savignii.*
D. *Echinoclinum verrilli, Eudistoma hepaticum, E. capsulatum.*
E. *Clavelina gigantea, Didemnum candidum, Trididemnum savignii.*
J. *Aplidium bermudae, A. exile.*
L. *Didemnum amethysteum, Trididemnum orbiculatum.*

By contrast, suspension feeders like Ascidians, living off shore on the gradually deepening bottoms off Georgia, are placed in a relatively warm, slow south to north ocean current. The Gulf Stream begins at the Straits of Florida and fans out northward mainly offshore. It passes off Hatteras and runs broadly northwest off the Atlantic coast. A weak north to south current passes close to shore off Georgia, but the coastal area is largely conditioned by slowly moving warmer water.

To a varying degree the two areas being compared are thus a northern counterclockwise vortex of cold water, and a slowly deepening southern shore current tending to be held at rather warmer temperatures. The Ascidian species of the Gulf of Maine are tough, well-established, tenacious sea squirts of the northern boreal area. The species of the Georgia coastal shelf are mainly species which lie loosely on the sand, or are attached together in masses of many species, living continuously on the food particles of a predominantly south-north current.

It is of interest to compare a preliminary list of the commoner Ascidian species of these two major sections of the Atlantic continental shelf. There is relatively little overlapping, although a few species can be found close to shore in both these continental shelf locations. The distribution of some of these species, restricted or wide-ranging, is contrasted in Figure 10.

Listed below are the commoner Ascidian species in two of the best studied sections:

GULF OF MAINE

Aplidium pallidum, A. glabrum, A. constellatum, A. stellatum, A. pellucidum
Ciona intestinalis
Distaplia clavata, Didemnum albidum, Trididemnum tenerum
Ascidia prunum, A. callosa, A. obliqua
Halocynthia pyriformis, Boltenia ovifera, B. echinata, Cratostigma singulare, Chelyosoma macleayanum

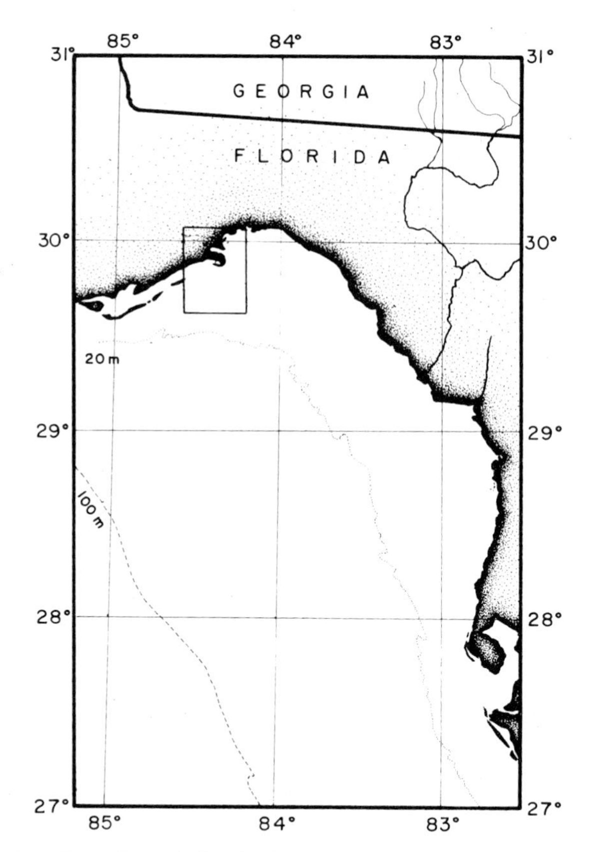

Fig. 8a. Location of Apalachee Bay, Florida.

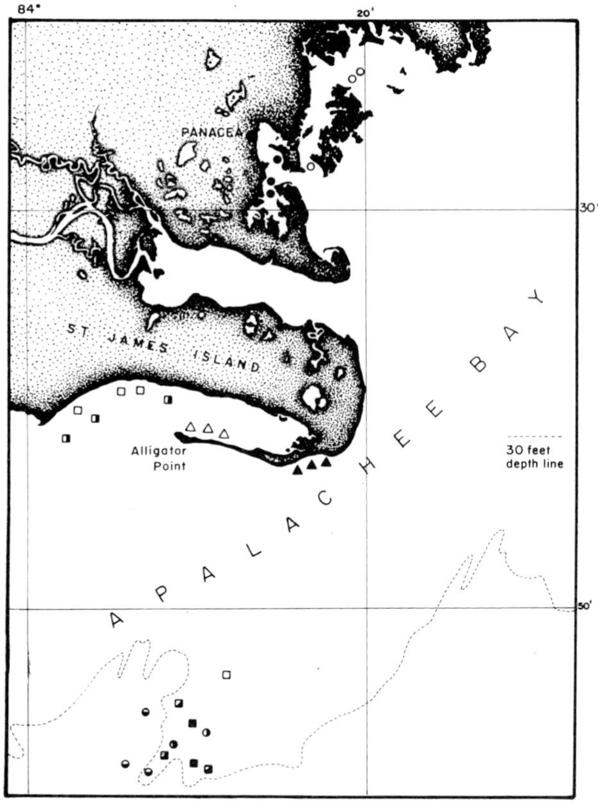

Fig. 8b. Apalachee Bay, Florida, and species collected there.

Perophora viridis (open circles)
Ecteinascidia turbinata (filled circles)
Distaplia bermudensis (open triangles)
Botryllus schlosseri (filled triangles)
Aplidium exile (open squares)
Polyandrocarpa maxima (diagonally half-filled squares)
Clavelina oblonga (vertically half-filled circles)
Eudistoma olivaceum (horizontally half-filled circles)
Didemnum candidum (vertically half-filled squares)
Polycarpa circumarata (filled squares)

Molgula siphonalis, M. retortiformis, M. citrina, M. complanata, M. manhattensis, M. arenata
Polycarpa fibrosa, Cnemidocarpa mollis, Styela partita, Dendrodoa carnea, Bostrichobranchus pilularis, Botryllus schlosseri

OFF SAPELO ISLAND, GEORGIA

Aplidium bermudae, A. exile, Distaplia bermudensis
Didemnum candidum, Trididemnum savignii, Clavelina oblonga, C. picta
Perophora viridis
Ascidia nigra, Pyura vittata, Molgula occidentalis, M. arenata, M. manhattensis
Styela plicata, Polycarpa obtecta, Symplegma viride, Botryllus schlosseri, Bostrichobranchus pilularis

It is apparent that the number of species is greater in the northern Atlantic region of the Atlantic continental shelf than it is in the Georgia region. Nevertheless, there are larger masses of Ascidian zooids growing together in the Georgia and Florida areas. Obviously, there are other factors than temperature and sea water currents which determine successful Ascidian species extension.

Similar but less extensive dredging was done on the continental shelf at several locations on the Florida west coast. Off Georgia the commonest species collected were *Styela plicata* and *Molgula arenata.* Off the west coast were found, at quite well separated depths and bottoms, *Polycarpa circumarata* and *Clavelina gigantea.*

In all areas there occur certain locations on appropriate marine bottoms which appear to stimulate the attachment of individuals of all the species having approximately similar living preferences. These have been called "live bottom habitats." In the Gulf of Maine such Ascidian-rich situations are found south of Mt. Desert Island, southeast of Cape Ann, and in the Vineyard Sound. There are similar favorable live bottom habitats in deeper water on Georges Bank. They can be found also inside the islands off Beaufort, North Carolina, about five miles east of Sapelo Island, Georgia, and off several of the Florida Keys, including Tortugas Island. On the west coast of Florida it appears that an especially favorable live bottom habitat is to be found in the shallow water of Apalachee Bay, south of Panacea, Florida.

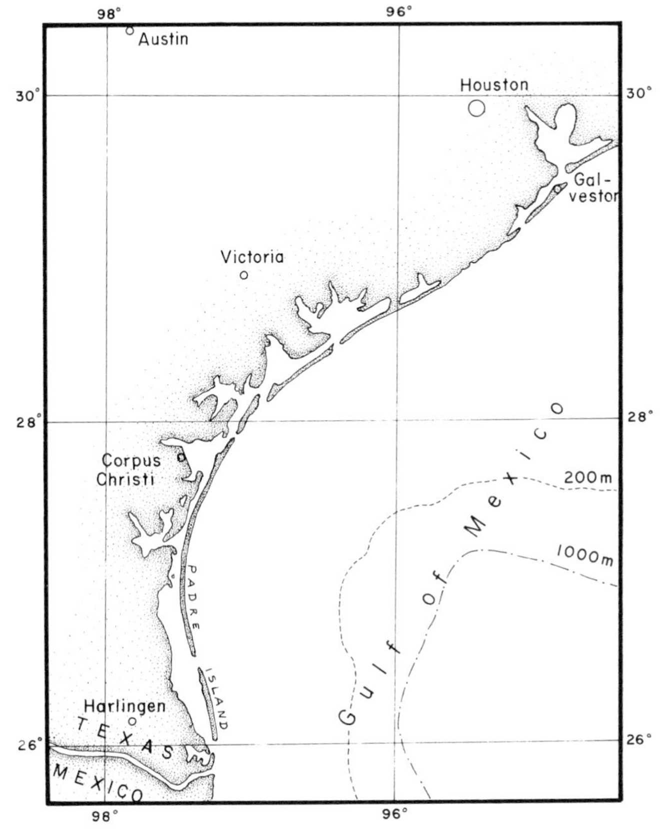

Fig. 9. Padre Island, Texas.

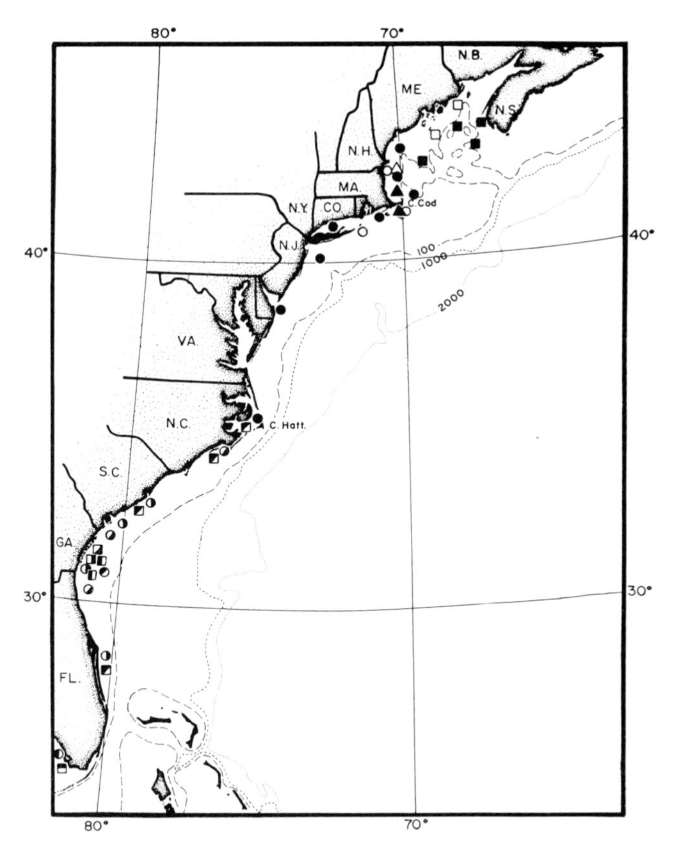

Fig. 10. Distribution of fifteen species off the east coast of the United States.

Aplidium glabrum (open squares)
Ascidia prunum (filled squares)
Chelyosoma macleayanum (open triangles)
Molgula citrina (filled triangles)
Dendrodoa carnea (open circles)
Ciona intestinalis (filled circles)
Aplidium bermudae (squares, diagonally cut, lower right filled)
Distaplia bermudensis (squares, vertically cut, right filled)
Ascidia nigra (circles, diagonally cut, lower right filled)
Molgula occidentalis (circles, vertically cut, right filled)
Styela plicata (squares, diagonally cut, upper left filled)
Clavelina oblonga (squares, vertically cut, left filled)
Trididemnum savignii (circles, diagonally cut, upper left filled)
Polyandrocarpa maxima (circles, vertically cut, left filled)
Polycarpa circumarata (squares, horizontally cut, top filled)

IV. Families of *Ascidiacea* Found on the Continental Shelf

The review of the Ascidians found on the Atlantic continental shelf begins with an examination of the major subdivisions used in classification, the *orders* and the *families*. Certain large families, in which a wide diversity is found, are further subdivided into *subfamilies*. These subfamilies within certain of the larger families often contain as many genera and species as some of the smaller families in the complete classification.

The important structural differences used in delimiting families have come to be: (a) the situation of the gonad within the body of the zooid and (b) the differentiation in the wall of the branchial sac. It is a striking fact that colony formers as well as solitary zooid species appear in every one of the suborders, so that these different results of vegetative budding have originated several times in the evolution of the class *Ascidiacea*.

The class *Ascidiacea* is divided into two orders, the *Enterogona* and the *Pleurogona*, based on the mode of formation of the atrial siphon and the position of the gonads. The order *Enterogona* is further subdivided into two suborders, *Aplousobranchia* and *Phlebobranchia*. The order *Pleurogona* has only one suborder in the Atlantic species, the *Stolidobranchia*.

As here subdivided there are twenty-four families or subfamilies of *Ascidiacea*, of which eighteen are found on the Atlantic continental shelf of North America, each of which shows a slightly different structural plan from the others. Most of these families appear in the northern boreal zone. As will be suggested later they may have migrated across the boreal zone or possibly they have simply been carried along with the continental splitting.

In Table I all of the families are listed, but those not found on the North American Atlantic shelf are enclosed in parentheses. The most useful arrangement of the Ascidian families seems to this author to be that proposed by Patricia Kott (1969) in her volume *Antarctic Ascidiacea*. Following an earlier suggestion of N. J. Berrill (1950), she has transferred the *Cionidae* from suborder *Phlebobranchia* to the *Aplousobranchia*, otherwise all colonial, thereby emphasizing the primitive character of the genus *Ciona* from which the colonial species were probably derived. Other changes correspond with arrangements of Berrill in his valuable volume *The Tunicata* (1950). This volume deals primarily with British *Tunicata*, but it follows several original

Table I. Key to the Orders and Families of *Ascidiacea*

Class *Ascidiacea*. Sac-like zooids with two siphons. Water current drawn through branchial slits by ciliary action.	A or B
A. Order *Enterogona*. Atrial opening formed by fusion of two lateral openings into atrial chamber. Gonads within loop of intestine.	C or D
B. Order *Pleurogona*. Atrial opening formed from single median dorsal invagination. Gonads on lateral body wall.	E
C. Suborder *Aplousobranchia*. Branchial sac without internal longitudinal vessels. Colony formers. Egg develops to complete tadpole larva.	
D. Suborder *Phlebobranchia*. Branchial sac without folds but with internal longitudinal vessels. Egg develops to complete tadpole.	
E. Suborder *Stolidobranchia*. Branchial sac with 4 to 7 folds (except Botryllids). Egg develops to reduced tadpole which may be anural.	

	Representative genus and species	Plate or figure number
Order *Enterogona*—Suborder *Aplousobranchia* Body divided into thorax, abdomen and sometimes post-abdomen. Neural ganglion between siphons above the neural gland. Gut loop posterior with heart at end of abdomen. Colonial except *Cionidae.*		
Family *Cionidae*—Solitary, five or more muscle bands on each side, gut loop posterior to branchial sac, many rows of stigmata, heart V-shaped, gonad inside gut loop. Embryonic epicardium present in adult as two lateral perivisceral sacs. Oviparous.	*Ciona intestinalis*	Plate I
(*Diazonidae*)—European, colonial species of moderately deep water in which zooids are divided into thorax and abdomen. They are oviparous. After the sexual cycle, each zooid strobilates and forms a colony, suggesting a relation to the structure in the next family.	(*Diazona violacea*)	
Clavelinidae—Colonial groups of zooids, partly independent, arranged in systems. Buds form stolonic vessels.		
Subfamily *Clavelininae*—Both siphons independent. Many rows of stigmata. Buds form statoblasts in winter.	*Clavelina oblonga*	Plate II
Polycitorinae—Branchial sac with reduced number of rows of stigmata. Budding by abdominal constriction.	*Eudistoma olivaceum*	Fig. 11
Holozoinae—Zooids in large masses arranged in irregular systems. Four rows of stigmata. Tadpoles develop in pocket extending into tunic. Buds form at tip of epicardium in abdomen.	*Distaplia bermudensis*	Plates III & IV
Polyclinidae		
(*Euherdmaniinae*)—(A Pacific species with two separate epicardia and two atrial siphons.)	(*Euherdmania claviformis*)	
Polyclininae—Compound species with massive colonies. Oral siphons separate, atrial siphons open into a common cloaca. Zooids long and divided into thorax, abdomen, and postabdomen. Complete swimming tadpole. Budding by constrictions of postabdomen.	*Aplidium constellatum* *A. pallidum*	Plates V, VI, & VIII
Didemnidae—Encrusting layers mostly white with calcareous spicules in test. Eggs break through into test and form tadpoles. Complex double pyloric buds unite to form one zooid.	*Didemnum albidum* *D. candidum* *Trididemnum savignii* *Diplosoma macdonaldi*	Plates VII & VIII
Enterogona—*Phlebobranchia* Body not divided into thorax and abdomen. Gut loop beside branchial sac, left or right. Heart a tube curving across branchial sac. Most species are solitary.		
Perophoridae—Branching colony with separate zooids on branching stolons, gut loop below on left side. Many rows of stigmata, which may be reduced to four. Viviparous.	*Perophora viridis* *Ecteinascidia tortugensis* *E. turbinata*	Figs. 12 & 13
Corellidae—Solitary, gut loop below and at the right side of branchial sac. Stigmata form spirals, gonads in gut loop. Dorsal lamina bears languets. Oviparous.		
Rhodosomatinae—Straight stigmata.	*Rhodosoma wigleii*	Fig. 14
Corellinae—Spiral stigmata.	*Corella borealis* *Chelyosoma macleayanum*	Fig. 14
(*Hypobythiidae*)—(Deep sea off South America)	(*Megalodicopia hians*)	

(Table I continued)	Representative genus and species	Plate or figure number
Ascidiidae—Solitary, gut loop on left side of branchial sac with many longitudinal vessels, renal vesicles with concretions. Oviparous or viviparous.	*Ascidia prunum*	Fig. 15
(*Agnesiidae*)—(Off tip of South America)	(*Agnesia glacialis*)	
(*Octacnemidae*)—(Deep sea mid-Atlantic)—(A rare disc-like species)	(*Octacnemus bythius*)	
Pleurogona—Stolidobranchia Body not divided. Gut loop on left side. Commonly four or more folds in branchial sac. Gonads on both sides of body. Neural ganglion usually below the neural gland. Eggs frequently form reduced tadpoles with only one sense organ or lose the tail, so are anural. Buds when formed are on lateral wall.		
Pyuridae—Solitary. Branchial sac with 5 or 6 folds. Stigmata usually straight. Complex hepatic diverticula. Mostly viviparous. Tadpoles usually complete. No buds.		
Bolteninae—Branchial stigmata transverse to body axis. Attached in moderate depths.	*Boltenia ovifera* *B. echinata*	Frontispiece & Plate IX
Pyurinae—Gonads many separate cases attached either side of oviduct. Shallow water.	*Pyura vittata*	Plate IX & Fig. 34a, b
Heterostigminae—Branchial folds absent. Spiral stigmata with flattened infundibula. Mostly shallow water.	*Cratostigma singulare*	Plate X
Molgulidae—Solitary. Siphons six lobed, branchial sac, usually with 7 folds. Stigmata spiral. Smooth stomach with hepatic glands. Renal sac on right side of branchial sac containing concretions. Oviparous or viviparous. Eggs develop into reduced tadpole larva with one sense organ only, or may be anural. No buds.		
Molgulinae—Infundibula well formed, under folds.	*Molgula citrina* *M. complanata* *M. manhattensis* *M. retortiformis* *M. arenata*	Plate XI Plate XII
Eugyrinae—Infundibula formed of two stigmata irregulargly distributed in branchial sac.	*Bostrichobranchus pilularis*	Plate XIII
(*Oligotreminae*)—(Deep sea—South Atlantic) Oral siphon modified for carnivorous diet.	(*Oligotrema psammites*)	
Styelidae—Solitary or colonial. Branchial sac with four folds, Stigmata straight. Stomach with longitudinal folds and, in addition, accessory hepatic vessels. No renal sac. Heart along endostyle bent to the right. Many gonads attached to body wall on both sides. Oviparous. Eggs form tadpoles with sense organs reduced to ocellus only.		
Botryllinae—Colonial. Zooid embedded in test. No folds. These are viviparous.	*Botryllus schlosseri*	Plate XIV
Polyzoinae—Colonial. Zooids bud from stolons.	*Symplegma viride* *Polyandrocarpa maxima*	Plate XV
Styelinae—Solitary. Wide variety of simple sac-like zooids adapted to many different marine habitats. Four branchial folds.	*Polycarpa fibrosa* *P. circumarata* *Cnemidocarpa mollis* *Styela partita* *S. plicata* *Dendrodoa carnea* *Dicarpa simplex*	Plate XVI & Figs. 16, 17, 18, 19, 20 & 21

studies of American shelf species, and it is illustrated with many original line drawings which apply equally well to species being discussed here. Several subfamily headings proposed by C. Monniot (1969) are adopted in the present review of Atlantic continental shelf species. In addition, the subfamily *Heterostigminae* is added here for the first time.

In the present review of Atlantic continental shelf families of *Ascidiacea*, the family *Styelidae* is placed last in the series, since this family includes a much wider range of distinctive generic lines than appear among the *Molgulidae* or the *Pyuridae*. These three families include the most advanced *Ascidiacea*, but the class has been in existence for a very long period of geologic time, and the *Styelidae* have made more experiments in structure and in the trial of different food currents than either of the other advanced families. The Styelids include *Styela* in shallow water, *Cnemidocarpa* farther out on the hard sand, *Polycarpa fibrosa* in deep mud pockets. They have experimented attached to rocks in strong current in the modified *Dendrodoa* and in colony formation by *Polyandrocarpa*. In addition, there are two major colonial subfamilies, *Polyzoinae* such as *Symplegma*, and *Botryllinae* with the wide spread *Botryllus* as examples of more extensive experimentation.

Table I charts the families and subfamilies of the whole class *Ascidiacea*. It is put in the form of a key which should make it possible to place a new specimen brought up in a collecting dredge in the correct family. This will require careful external examination, and later careful dissection with fine scissors. Any initial difficulties will require one or more additional specimens for study. If only one specimen is secured, it may be more satisfactory to narcotize it with epsom salts in the sea water, then preserve and harden it, when it will be possible to dissect without destroying it. One or more common species illustrating each family found on the Atlantic continental shelf are shown in the plates or the subsequent figures. In the exceptional case of the few species which show no bright color in life, the species is drawn as in transparent view in accurate black line sketches.

The key (Table I) gives the distinguishing characters of each family and the plates and figures show representative species. Typical but easily available species have been selected for illustration of each family, and sometimes more than one species has been pictured. It is hoped that the method of presenting pictures of representative species of every family and subfamily will make the final identification of newly collected specimens easier. One can locate them first in the correct family, and later work out the species within that family.

The listing of the families and subfamilies of *Ascidiacea* in Table I shows that most of the families of *Ascidiacea* of the world are represented on the Atlantic continental shelf. Only six of the family groups listed have failed to appear in the sporadic collecting of the past one hundred years. Two of these families, *Diazonidae* and *Euherdmaniinae*, represent two interesting evolutionary links which failed to reach the American continental shelf. In *Diazona* there are two separate epicardia, and it is a colonial species of European seas, perhaps related to *Ciona*. In contrast, *Euherdmania* is a Pacific shelf species, solitary like *Ciona* and also possessing the primitive two separate epicardia.

The other four families not represented on the North American Atlantic shelf are deep sea dwellers and have a wide distribution in the depths of many oceans. Probably more extensive collecting from deep areas like the Bermuda trench may eventually show that some of these highly adapted deep sea Ascidians live also on the continental slope east of the United States. The families are *Hypobythiidae*, *Agnesiidae*, *Oligotreminae*, and *Octacne midae*, all of which have been taken from deep water in the South Atlantic. Representative species from each family suggest that the basic Ascidian stock has been capable of giving rise to some bizarre lines with unexpected evolutionary potentials.

In the antarctic *Hypobythiidae* a genus called *Megalodicopia* is a large sac attached to the bottom with spreading wing-like siphons and a branchial sac with very irregular stigmata. It is related to the *Corellidae* and shows that deep sea species may have developed their own peculiar collecting methods. Another widely distributed antarctic family, *Agnesiidae*, shows spiral stigmata and is a second evolutionary line derived from the family *Corellidae*. *Agnesia glacialis* is widely spread in the Pacific as well as in the South Atlantic. All of these strange deep water Ascidian species are illustrated in Patricia Kott's *Antarctic Ascidiacea*. The present survey has shown several species of the ancient family *Corellidae* living in the Gulf of Maine. Although none of them are often found, their wide distribution in all seas, including deep sea bottoms, suggests that this family has been an ancient root family competing with *Ascidiidae*.

Finally, there is an interesting Molgulid subfamily represented by the species *Oligotrema psammites*, found in antarctic deep water north of the South Shetland Island, as well as in the Indian Ocean, and also north of Iceland. It shows a reduced branchial sac and anterior muscular walls. In this genus the Ascidian method of suspension feeding has been largely superseded by muscular folds about the oral siphon by which jaw-like processes draw in larger food particles. Thus a carnivorous method of feeding is initiated. Although not found in the Atlantic shelf collections, these bizarre antarctic Ascidians clearly show some of the interesting

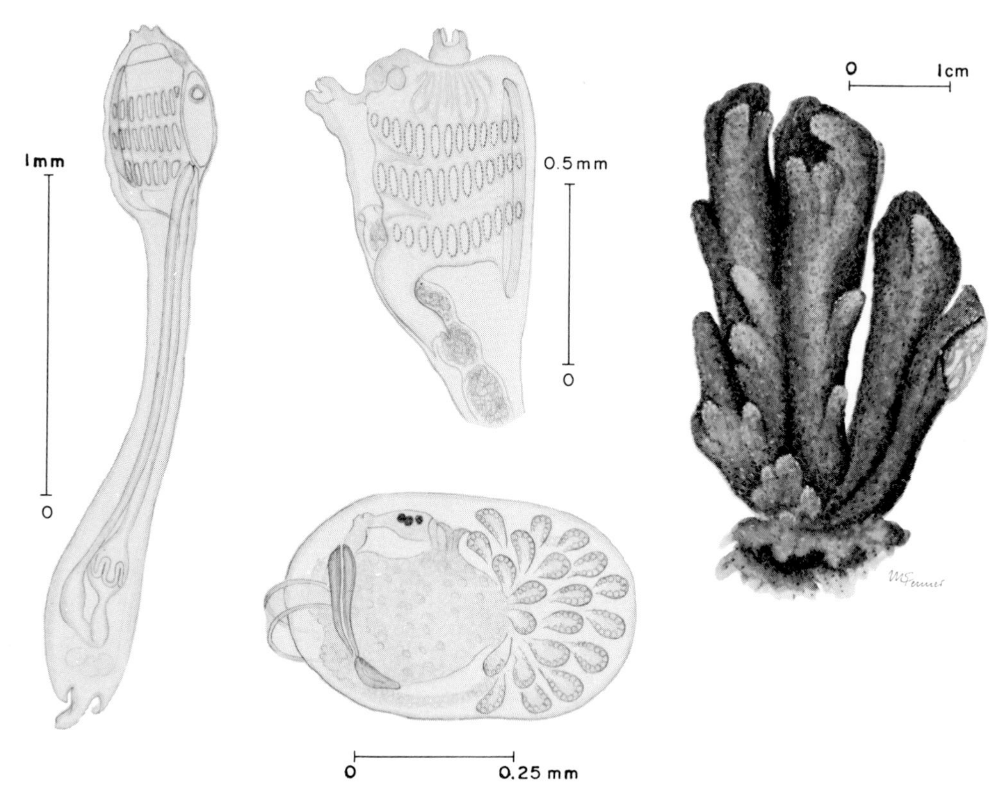

Fig. 11. *Eudistoma olivaceum.* Young adult (left), young bud (upper center), tadpole (lower center), colony (right).

evolutionary experiments of the *Ascidiacea* in the long past.

A family known as *Octacnemidae* seems to fit between the *Ascidiacea* and the pelagic *Tunicata*, the *Thaliacea.* Specimens from deep sea dredging off Patagonia studied by Metcalf (1895) showed an eight-pointed attached zooid, suggesting a common pattern with the free-swimming *Thaliacea.* Again the survival of Ascidian species which may represent evolutionary interconnecting links with more successful taxonomic groups is emphasized.

There is no better way to visualize the place of the class *Ascidiacea* as a world-wide group of solitary or colonial benthic marine animals than to review the families found on the Atlantic continental shelf. This can be done best by following the key in Table I. By carrying a new specimen through the alternative choices its family and often even its species name can be determined.

The order *Enterogona* is made up of two suborders: *Aplousobranchia* and *Phlebobranchia.*

Aplousobranchia include six families or subfamilies, all except the first forming massive colonies in shallow water or at medium depths. Their colonies are bulbous or thinly encrusting. They are often bright colored when alive and they may form thick masses overgrowing rocks, molluscs, or coral. It is possible that these colonial families, along with *Cionidae*, are the most ancient Ascidians. All are colonial except *Ciona* (Plate I) which may have colonial descendants. In all these *Aplousobranchia* the egg gives rise to a typical swimming tadpole, bearing two sense organs, an eye spot and an otolith. They differ greatly in vegetative bud formation, which in all cases starts with a thickening of the posterior body wall. In *Clavelininae* (Plate II), e.g., *Clavelina oblonga* & *C. gigantea*, with two separated siphons, buds are formed from vascular tissues. In *Polycitorinae*, e.g., *Eudistoma olivaceum* (Fig. 11), buds arise by layering or strobilization of the abdomen as they do in *Holozoinae* (*Distaplia bermudensis*, Plates III & IV). In *Polyclininae*, e.g., *Aplidium constellatum* (Plates V & VI), buds are from undifferentiated tissue masses in the abdomen. In *Didemnidae* unusual double buds are formed at the base of the branchial sac, which join together to form a complete adult zooid (Plates VII & VIII, *Didemnum albidum*, *D. candidum*, *Tridemnum savignii*, and *Diplosoma macdonaldi*).

For all their differences in colony appearance, all of the *Aplousobranchia* show basic resemblances in tad-

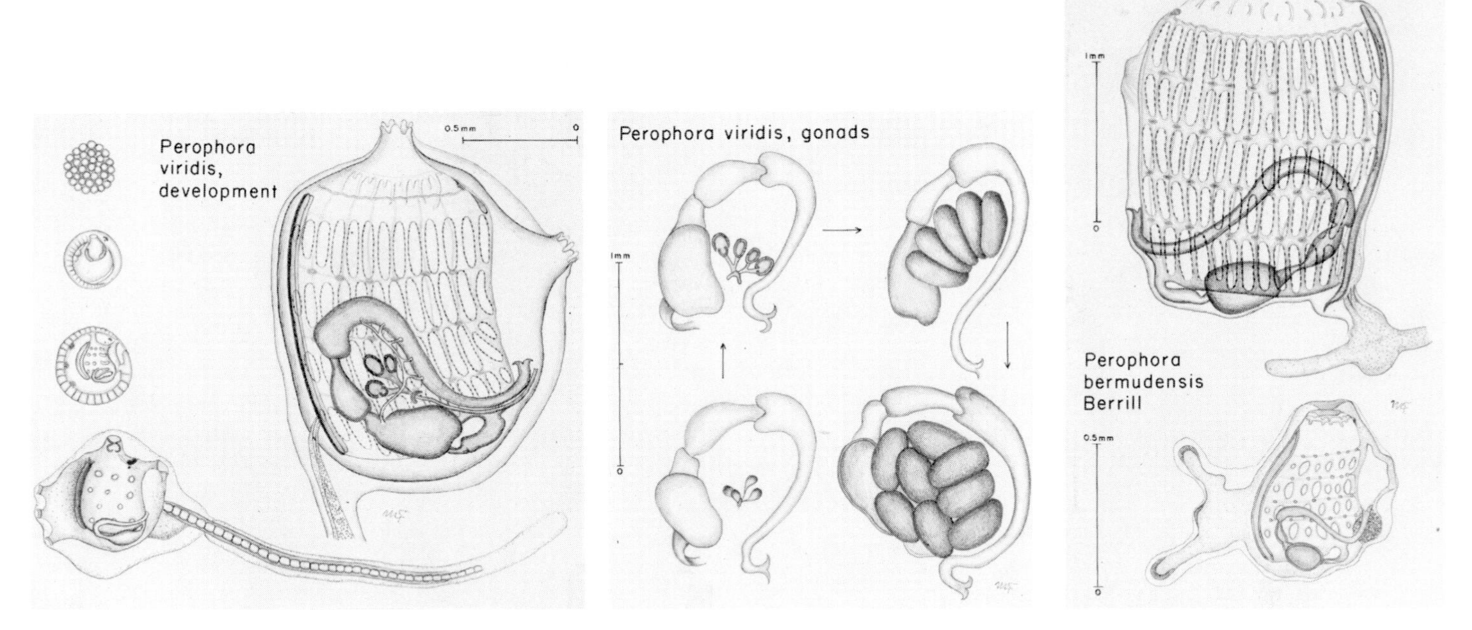

Fig. 12. *Perophora viridis*: adult, detail of gonads, development. *P. bermudensis*: adult. (After Berrill)

pole development and bud formation. With *Ciona* they were the earliest of the *Ascidiacea*. The *Polyclinidae* and *Didemnidae* are most numerous in the moderately deeper water in the Gulf of Maine, the northeast portion of the Atlantic shelf, but are found in shallow water in the Gulf of Mexico. The *Cionidae* occur frequently in harbors everywhere, both in the northeast and occasionally farther south. The three subfamilies of *Clavelinidae* are found, in contrast, only in the southern portion, especially off Georgia and Florida. So the earliest ancestors of the *Aplousobranchia* could have come to the North American Atlantic continental shelf from the boreal north, or with continental separation could simply have divided the area of their original distribution.

The *Phlebobranchia* include only three families on the Atlantic continental shelf. The first is *Perophoridae*, which includes *Perophora viridis* (Fig. 12), in the northern portion and several southern more massive colonial species like *Ecteinascidia turbinata* (Fig. 13). The Perophorids have small zooids growing on a tiny treelike stalk in shallow water, with algae mixed in. The southern Ecteinascidians grow encrusting colonies and consist of larger individuals attached to basal encrusting stolonic masses. All zooids are interconnected by stolons bearing branching blood vessels.

Both the other families of *Phlebobranchia* consist of solitary species. One of these is the widely distributed but never common family of *Corellidae*. This is an ancient stock widely distributed but never found in large numbers. Both subfamilies, the *Rhodosomatinae* and *Corellinae* (Fig. 14), have been dredged from off Cape Ann in the Gulf of Maine, suggesting preference for the cold subarctic current.

The other major ancient family, the *Ascidiidae*, contains many important widespread and distinctive species. It includes *Ascidia prunum* in the cold water group which spreads at moderate depths throughout the Gulf of Maine (cf. Fig. 15). In addition, another series of species of the family *Ascidiidae*, as *Ascidia nigra*, is found distributed close to shore on sand or stones in the southeast and off the West Indies. It is often found on mangrove roots. These three families include Ascidian species which after *Ciona* are among the oldest solitary protochordates on the North American eastern continental shelf.

The order *Pleurogona* and suborder *Stolidobranchia* are identical in present usage. The suborder heading is retained only to indicate a closer interrelationship within the major families to the just cited *Ascidiidae*. There are three major families in the *Pleurogona* subdivided into nine subfamilies. They include the most highly differentiated of living *Ascidiacea*. All of them show folds in the branchial sac, but in a few genera the folds have been lost secondarily. A number of structural specializations have developed, and some have modified or even discarded the tadpole as a developmental stage.

Among the three subfamilies of the family *Pyuridae*

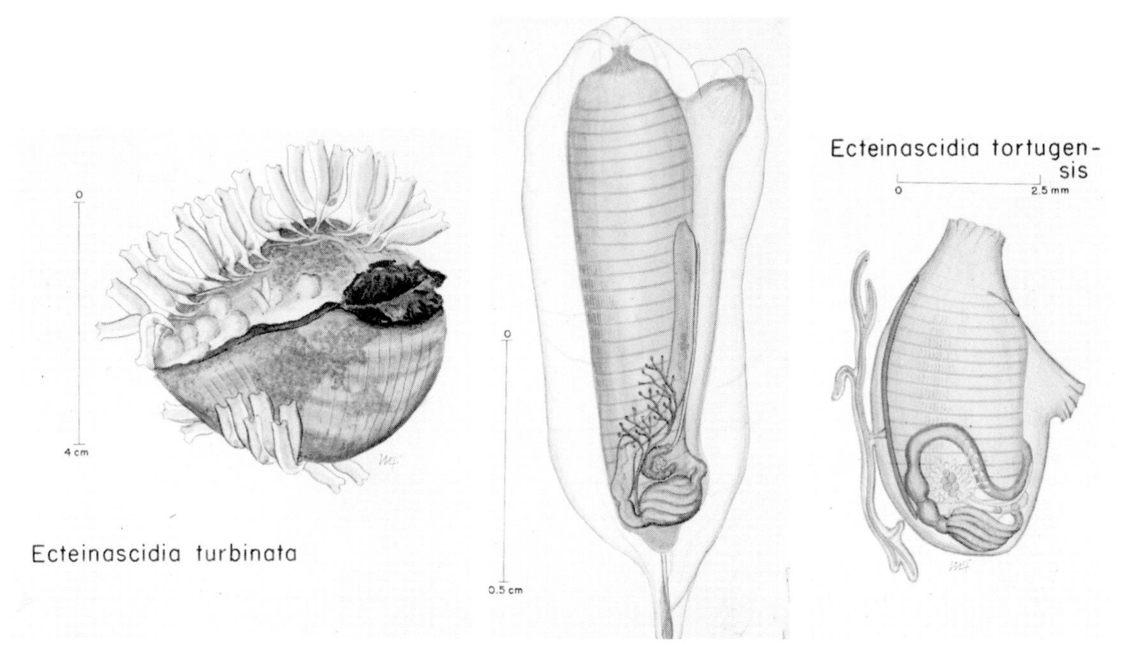

Fig. 13. *Ecteinascidia turbinata*: multiple growth on a shell, individual. *E. tortugensis*: adult.

there are several unusual specializations, such as long stalks for solid attachment in deeper water, and transversely placed stigmata (Cf. *Boltenia ovifera*, frontispiece), instead of the longitudinal arrangement in all other Ascidian genera. Yet the commonest species, *Boltenia echinata* (Plate IX), found in deeper water throughout the Gulf of Maine forms a typical tadpole larva with the original two sense organs. In one curiously unusual subfamily, the *Heterostigminae*, living north and south of Cape Cod and out into deeper water on Georges Bank, there are well-developed spiral stigmata in the branchial sac (cf. Plate X, *Cratostigma singulare*).

The family *Molgulidae* has evolved along a narrow evolutionary pathway. There are six to seven folds in the branchial sac, and the stigmata are always at least partly spiralized. Their structure is well shown in *Molgula citrina* (Plate XI) and *Molgula complanata*, each with a tadpole which has only one sense organ. Molgulid species like *Molgula manhattensis* are better known to marine zoologists because they live close to shore. Certain others like *Molgula arenata* are found in large numbers on hard sand up to several miles at sea off the shore of Georgia (cf. Plate XII). Still other active species secured in Cape Cod Bay and off shore in the Gulf of Maine, like *Molgula retortiformis*, survive because they are supplied with ample food particles by active marine currents.

A most unusual and widely successful Molgulid is found in mud pockets in two fathoms of water here and there up and down the Atlantic shelf. It is *Bostrichobranchus pilularis* (Plate XIII), a small Ascidian in which the branchial sac has lost the folds and instead has the stigmata elongated, spiralized and extended off the surface. These nipple-like cones, called infundibula, greatly increase the branchial sac surface for oxygen interchange as well as the picking up of food particles. The species is found all the way from north of Cape Cod to off Padre Island south of Corpus Christi, Texas.

It is the family *Styelidae* which appears to have tried the greatest number of divergent structural experiments among the *Pleurogona*. It includes three distinct subfamilies which must have had an ancient origin and which are very widely distributed on warmer continental shelves around the world. The best known subfamily is the *Botryllinae*, which are bright-colored masses of tiny zooids with separate oral siphons and united with six others about one common atrial siphon. The encrusting masses are brown, or pink, or blue, or purple, or a combination of these. The apparent star-shaped colonies make mats which are purple or gold. They are found in moderate or warmer latitudes and are often the most distinctive animal patches along shallow ocean wharf piles. Plate XIV shows a typical red *Botryllus* colony. During the summer, glass slides placed in wharf boxes will show attached tadpoles which develop in a day or two into colonies for study.

Ciona intestinalis, both sides, parts of oviduct, sperm duct

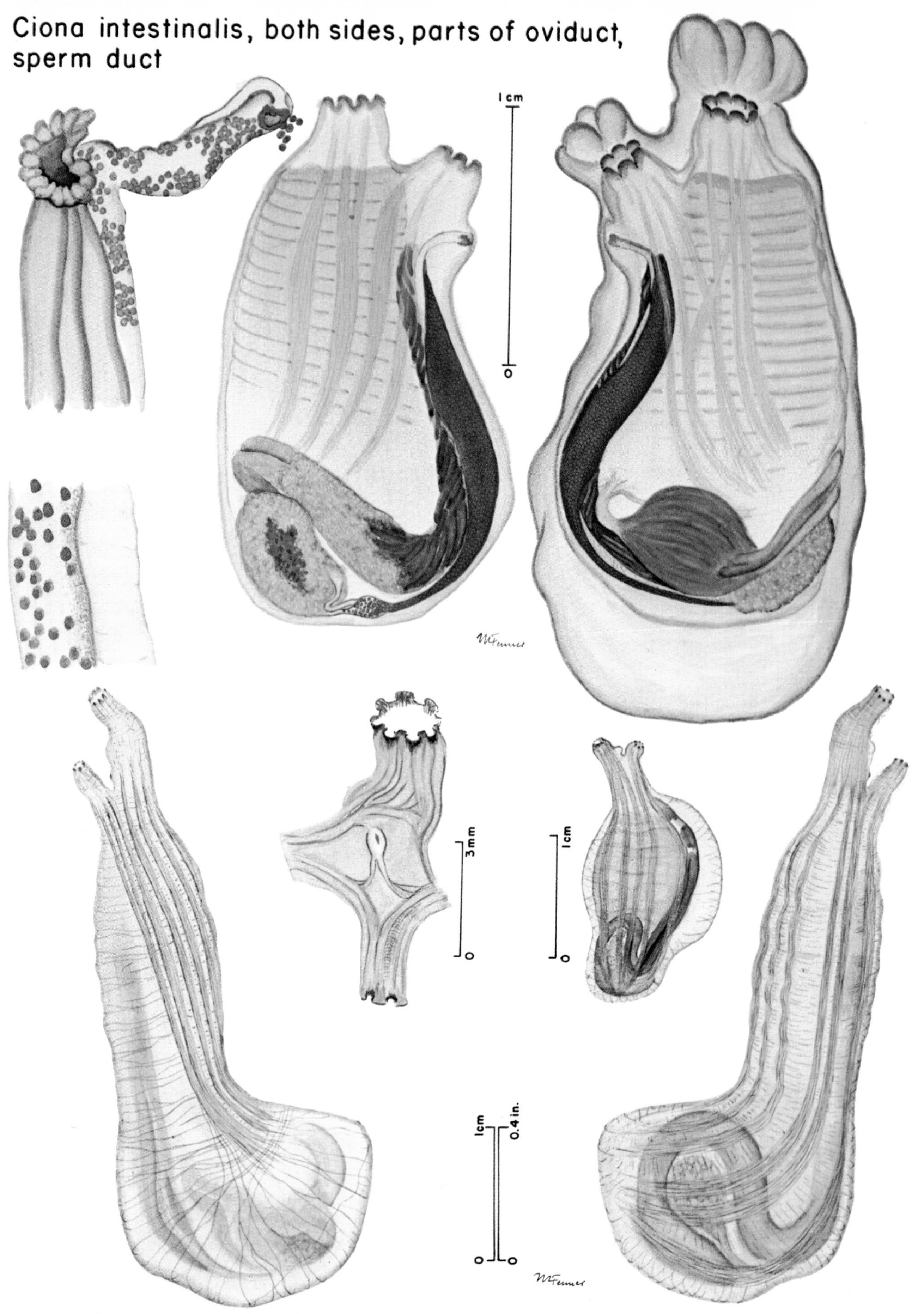

Ciona intestinalis, large and young animal, dors. tubercle

PLATE I

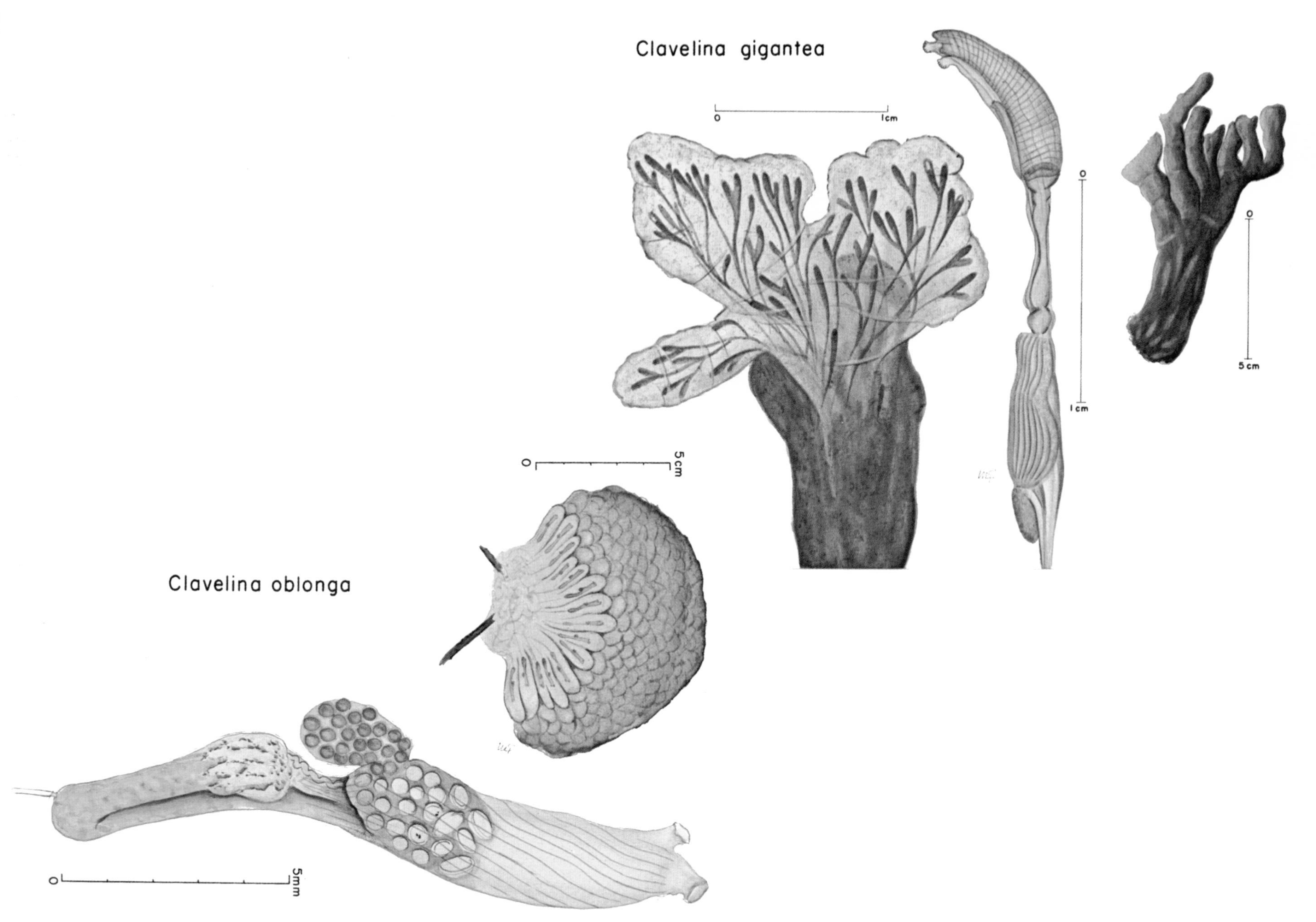
Clavelina gigantea
Clavelina oblonga
0
1cm
5cm
5mm

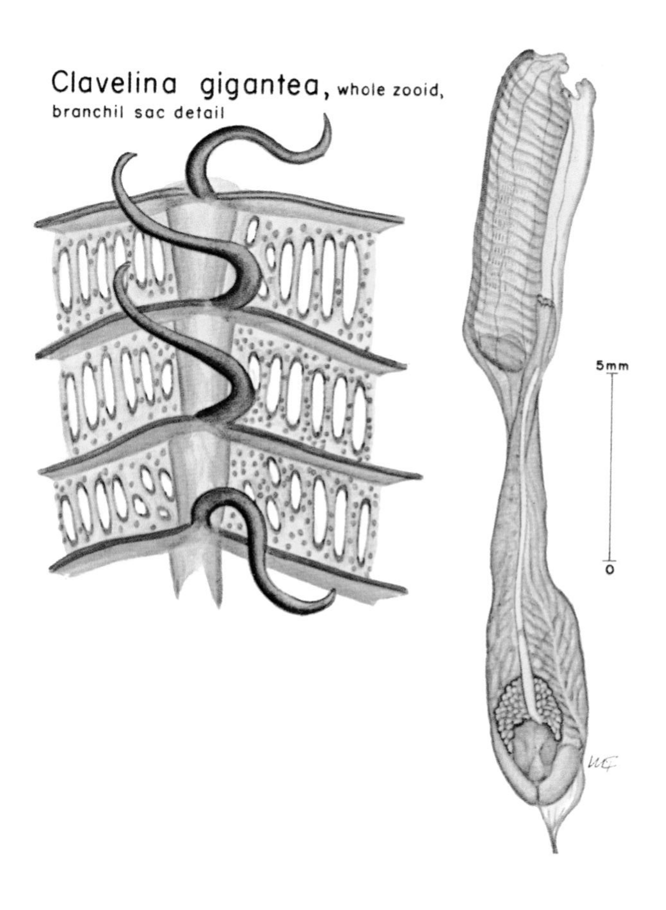
Clavelina gigantea, whole zooid,
branchil sac detail
5mm
0

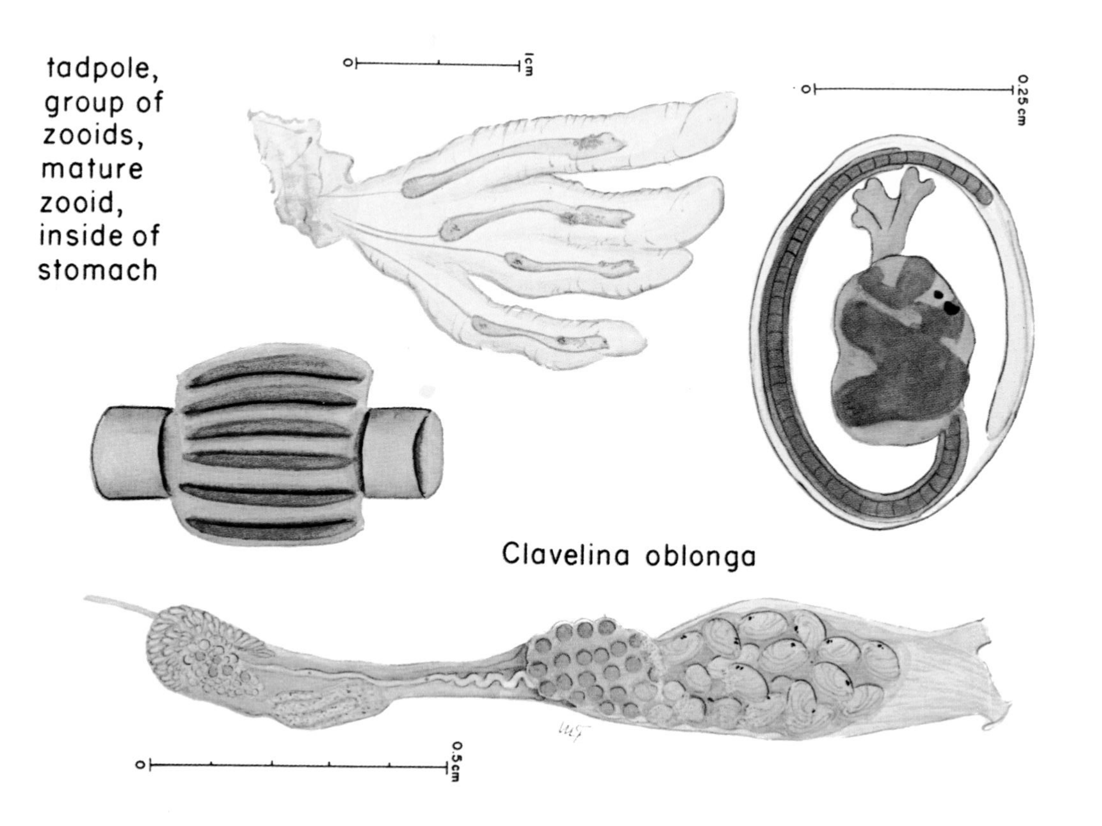
tadpole,
group of
zooids,
mature
zooid,
inside of
stomach
1cm
0
0.25cm
0
Clavelina oblonga
0.5cm
0

PLATE II

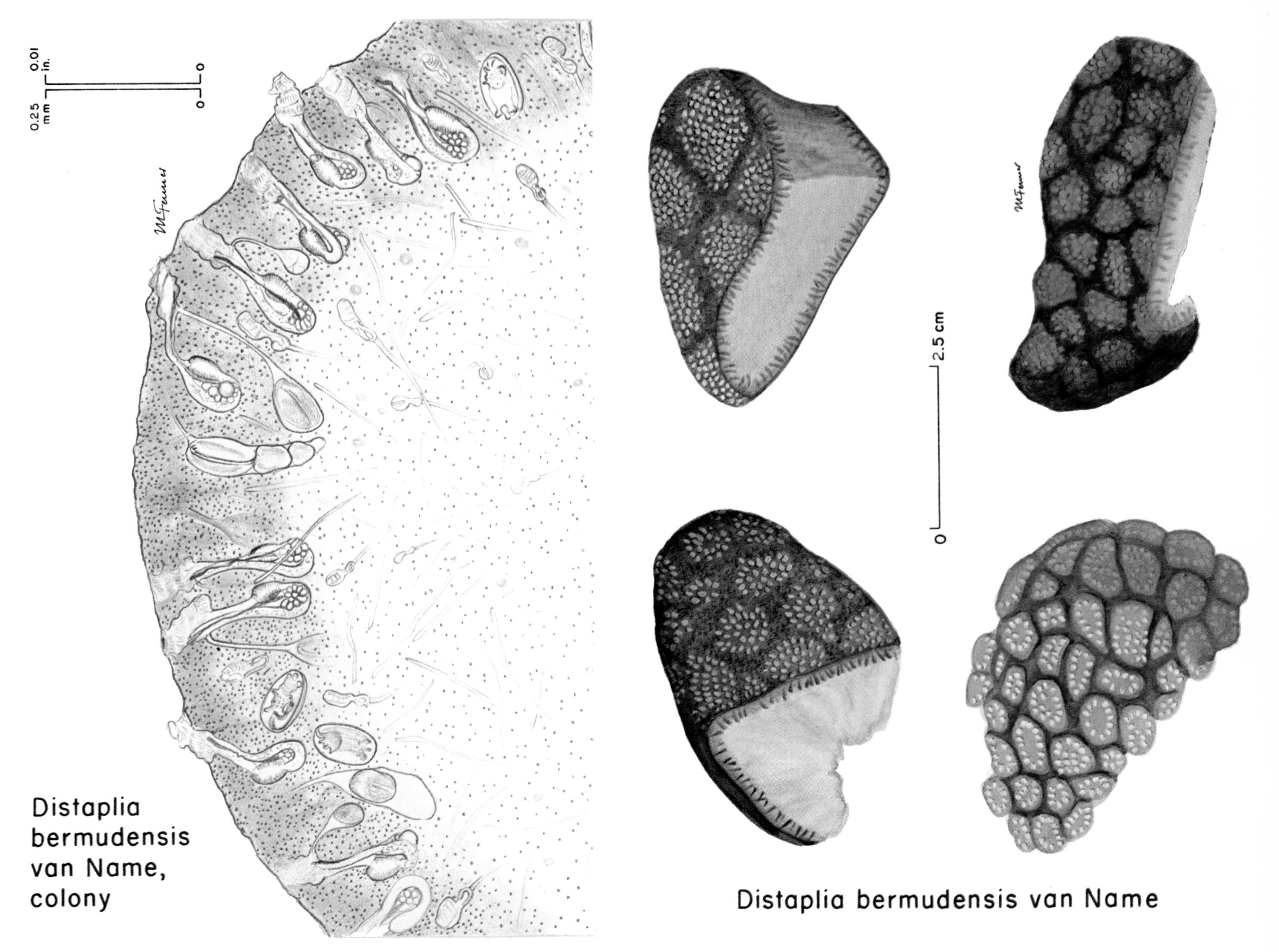

Distaplia bermudensis van Name, colony

Distaplia bermudensis van Name

PLATE III

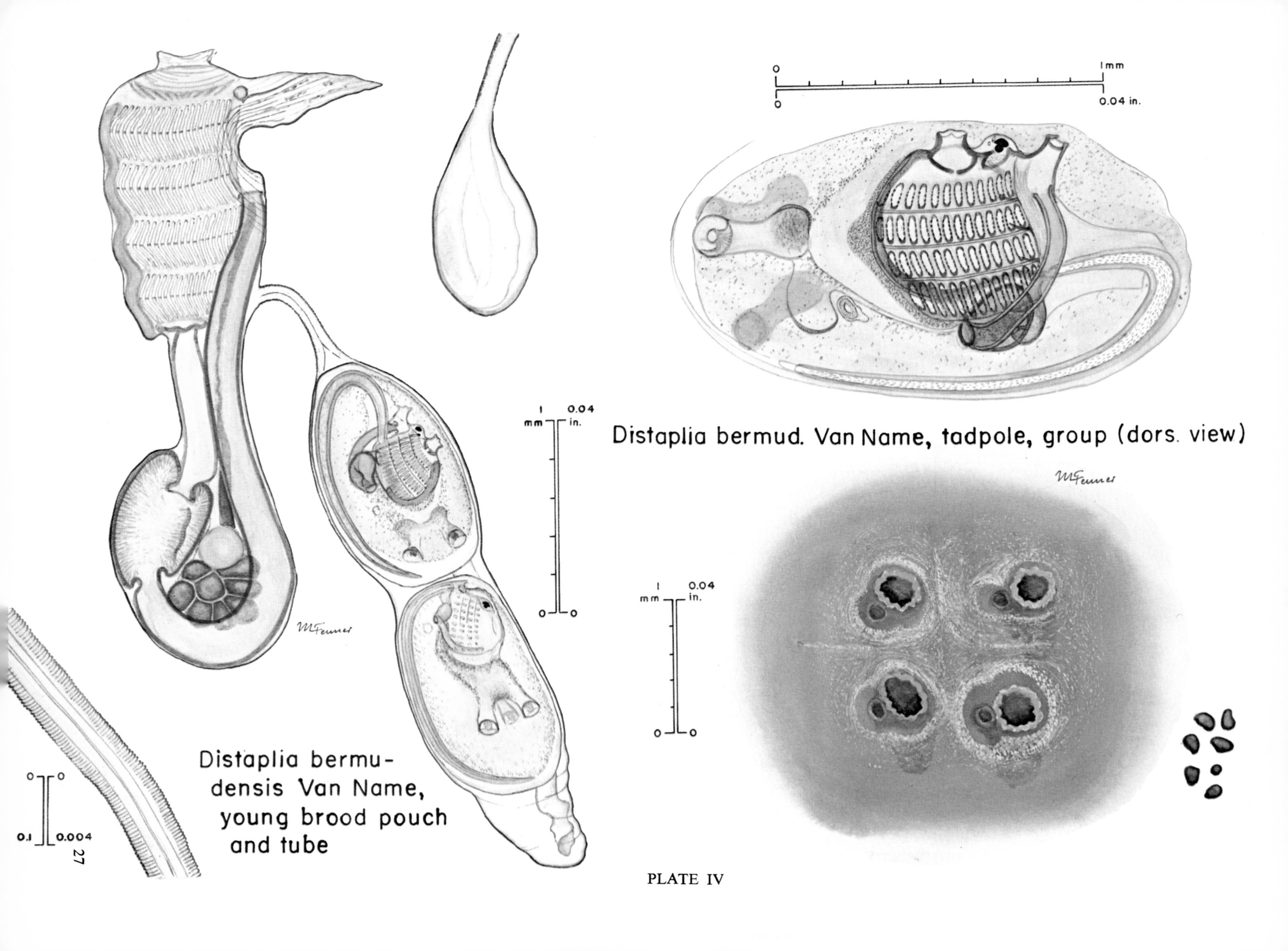

Distaplia bermud. Van Name, tadpole, group (dors. view)

Distaplia bermudensis Van Name, young brood pouch and tube

PLATE IV

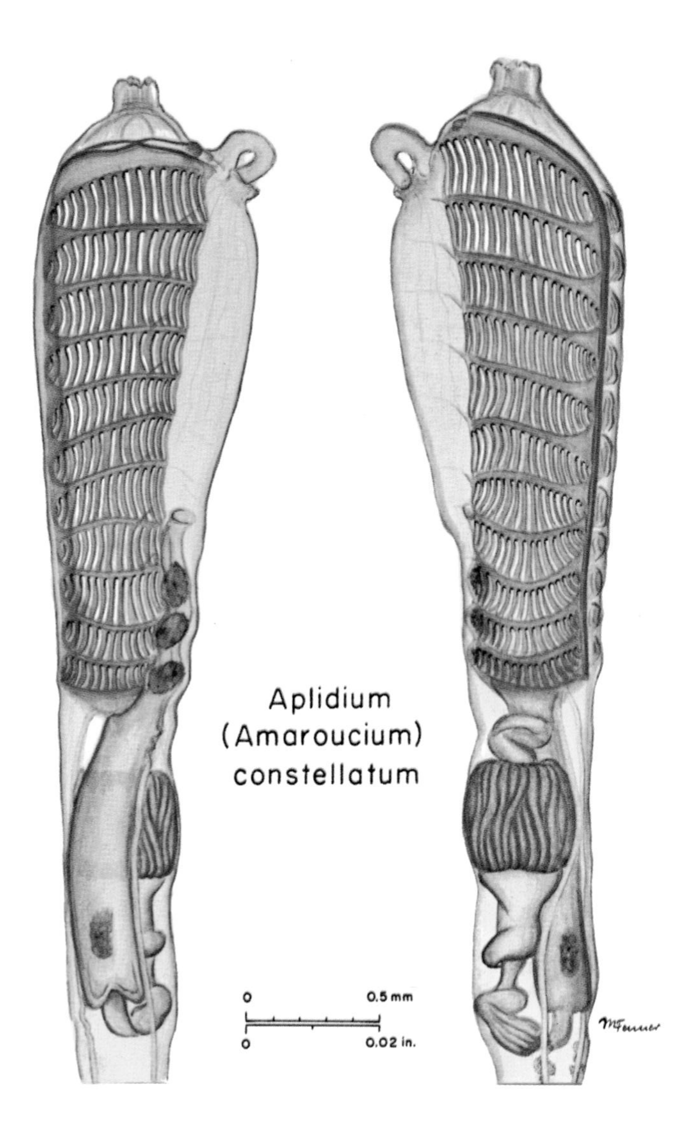
Aplidium
(Amaroucium)
constellatum
0 0.5 mm
0 0.02 in.

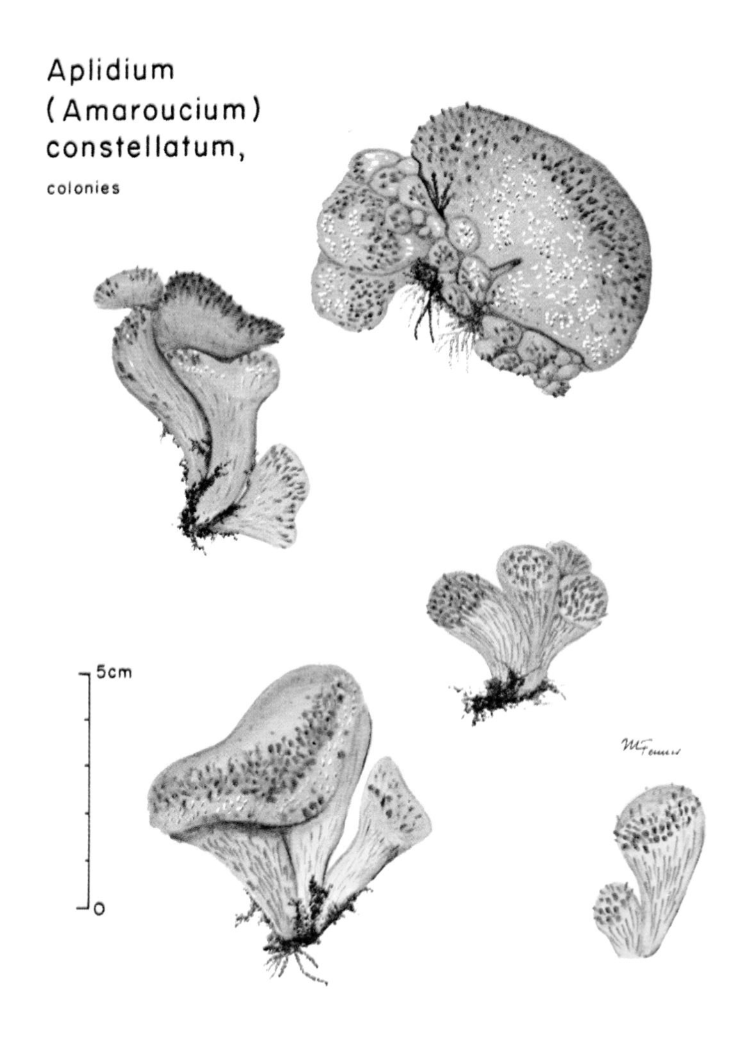
Aplidium
(Amaroucium)
constellatum,
colonies
5cm
0

PLATE V

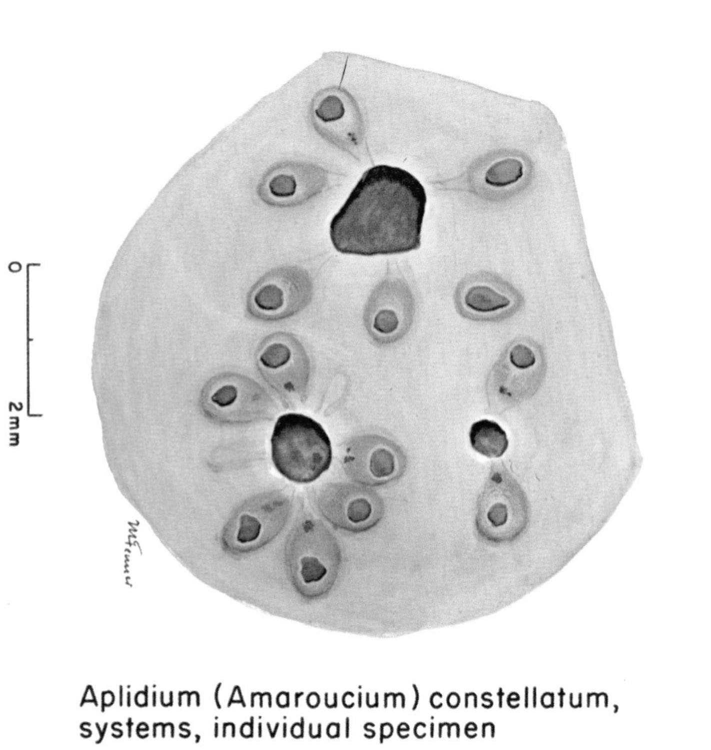

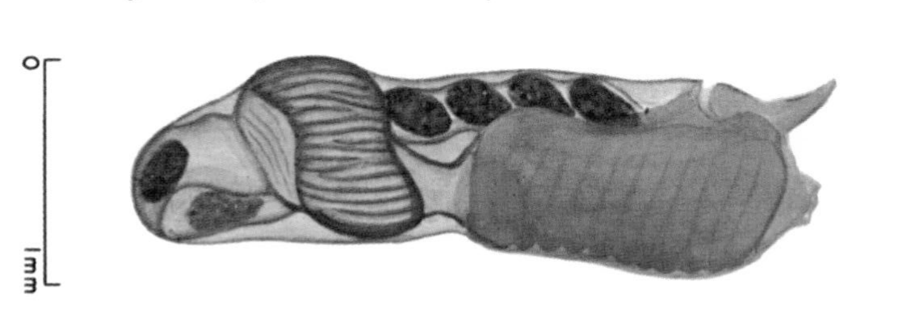

Aplidium (Amaroucium) constellatum, systems, individual specimen

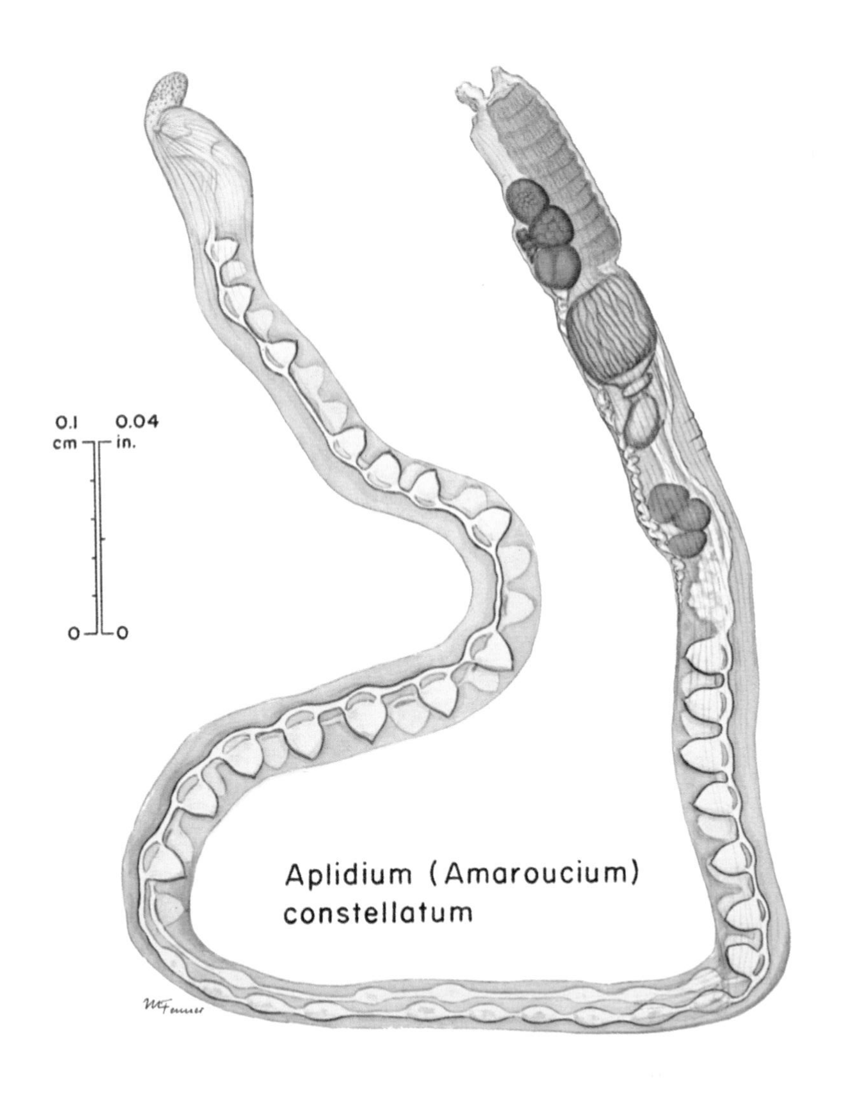

Aplidium (Amaroucium) constellatum

PLATE VI

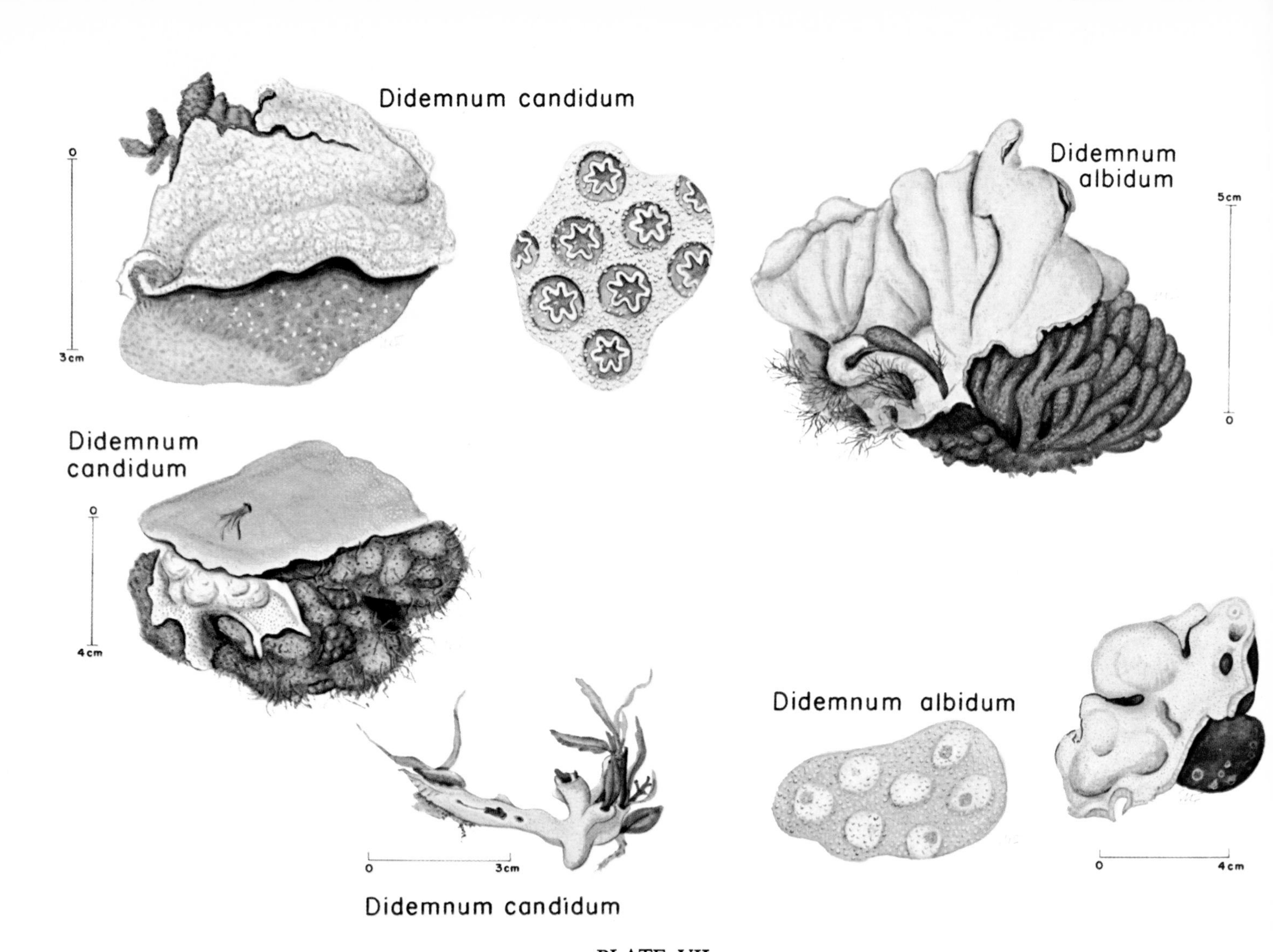

PLATE VII

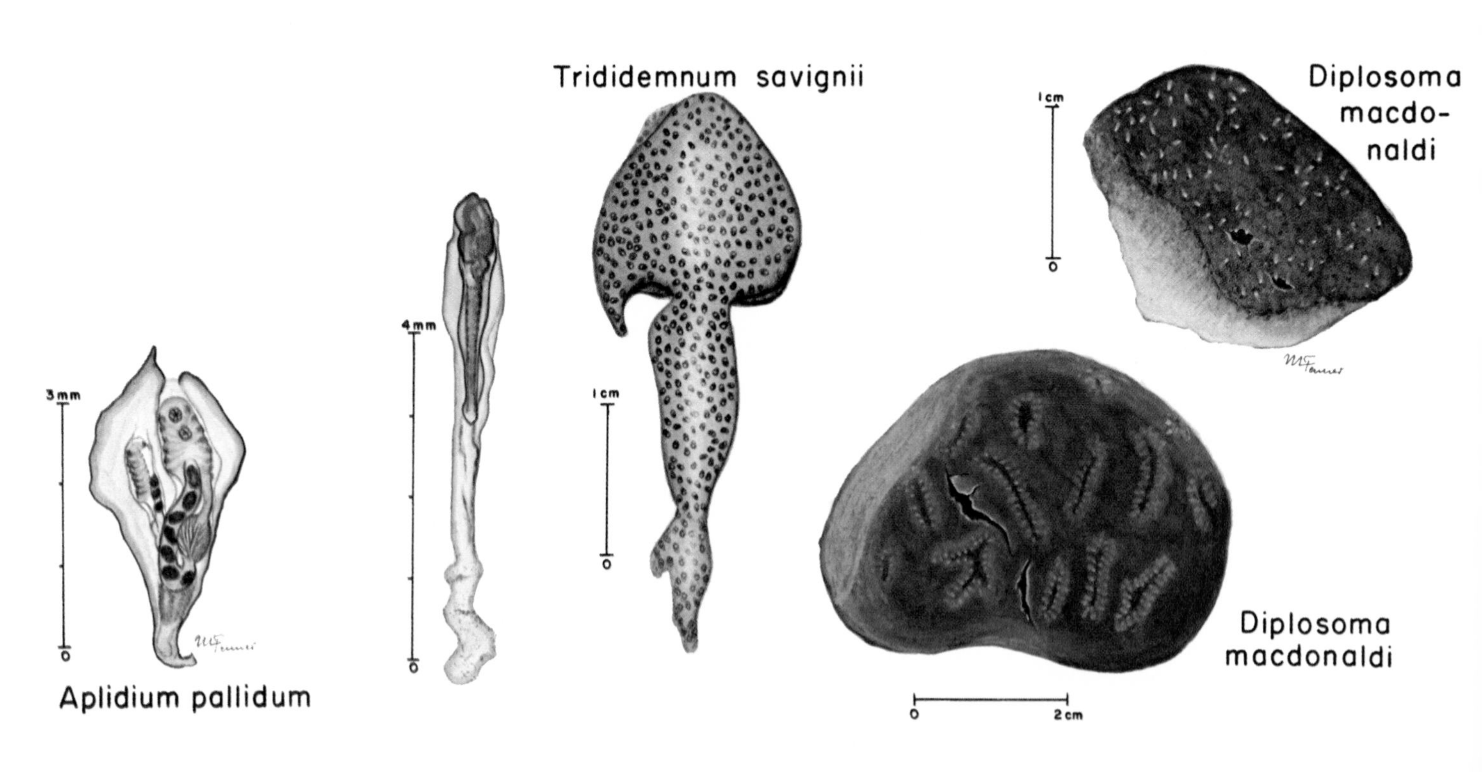

PLATE VIII

Boltenia echinata

Pyura vittata

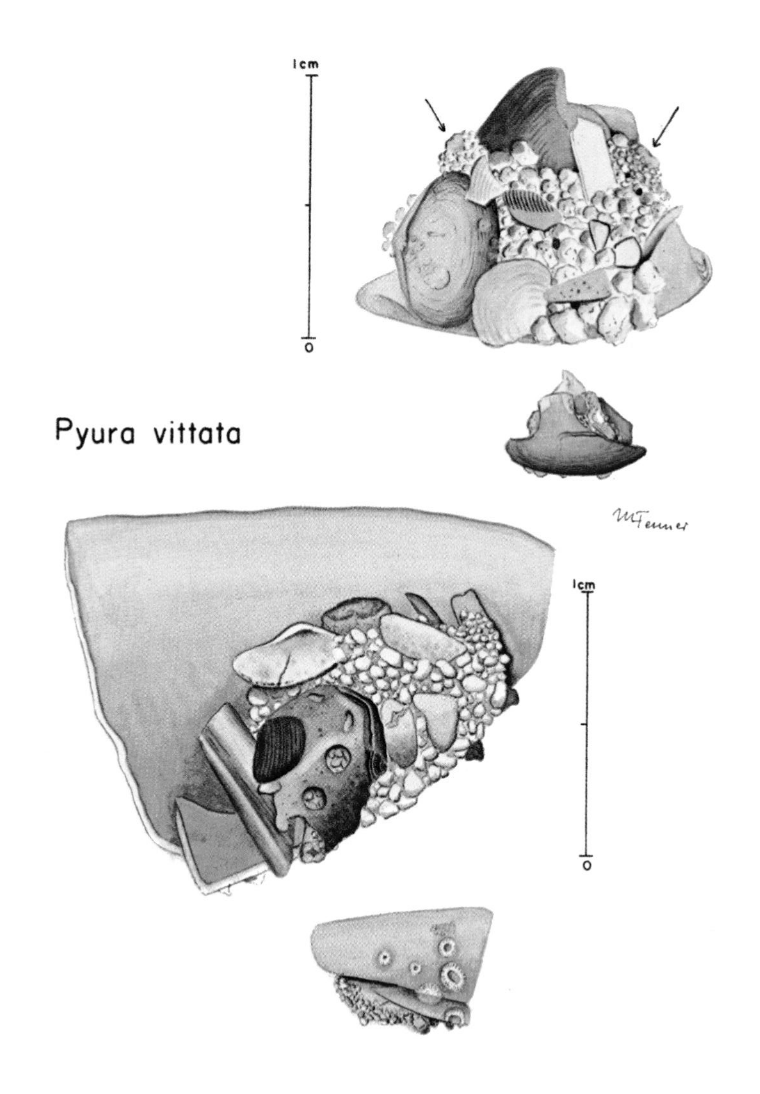

PLATE IX

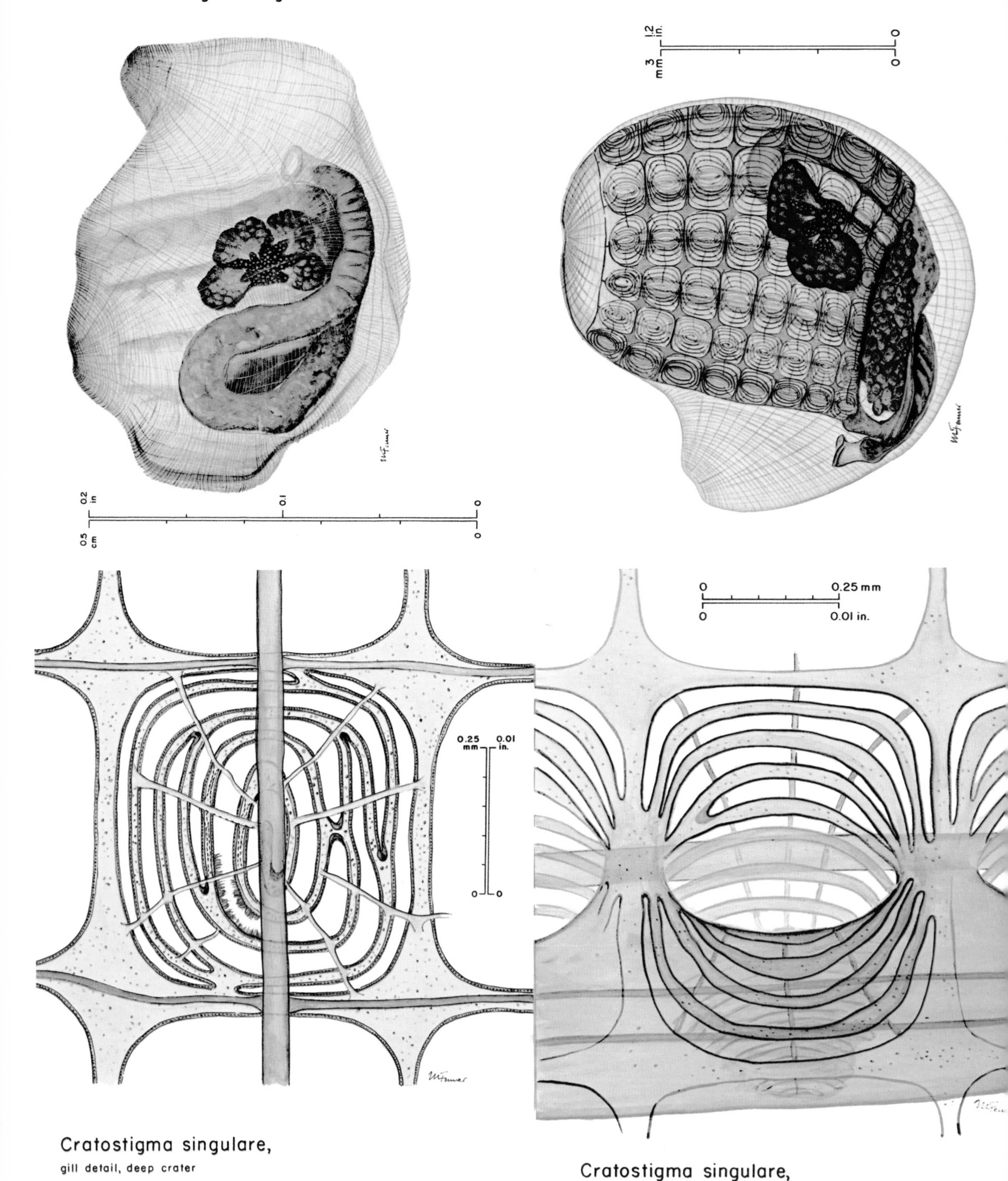

Cratostigma singulare

Cratostigma singulare, gill detail

Cratostigma singulare, gill detail, deep crater

Cratostigma singulare, gill detail, shallow crater

PLATE X

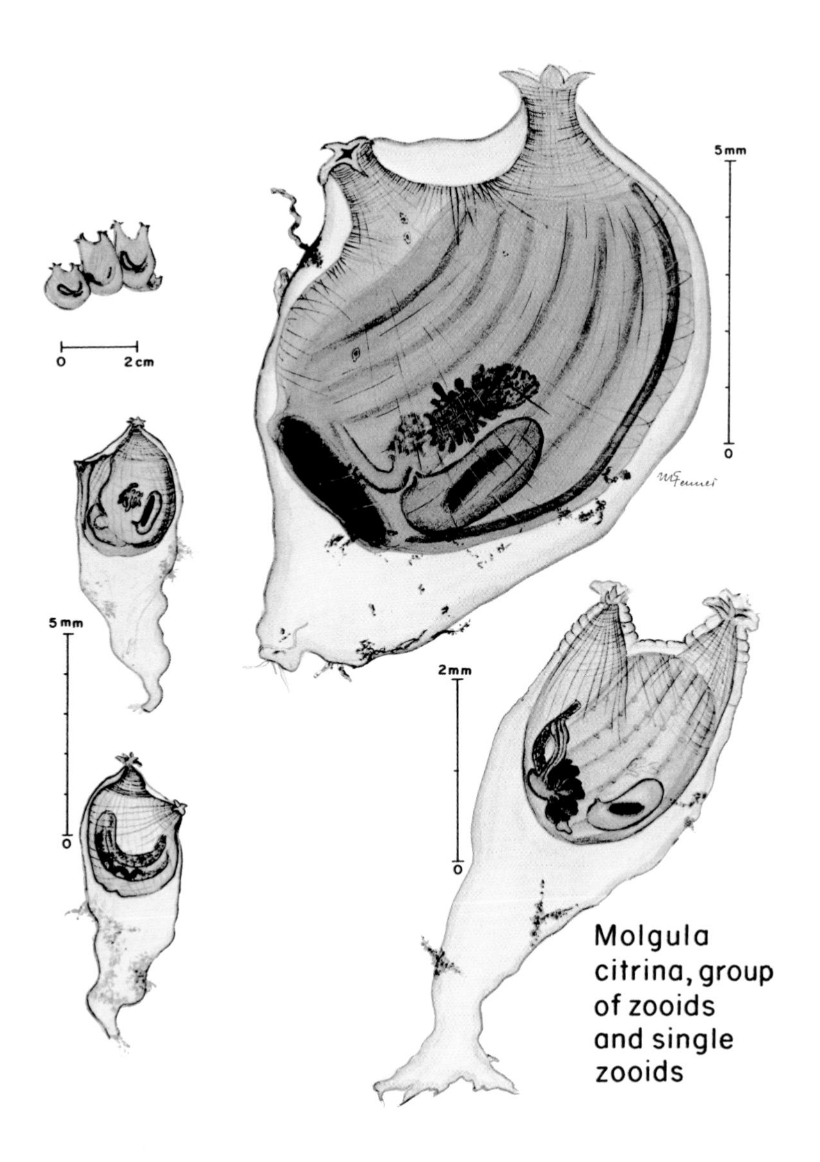

Molgula complanata

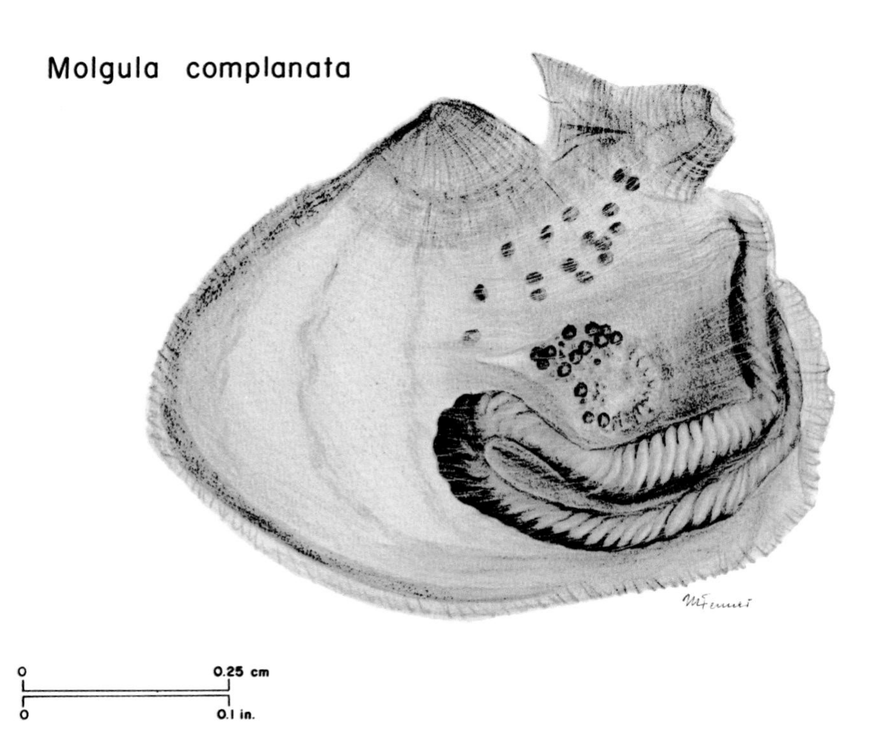

PLATE XI

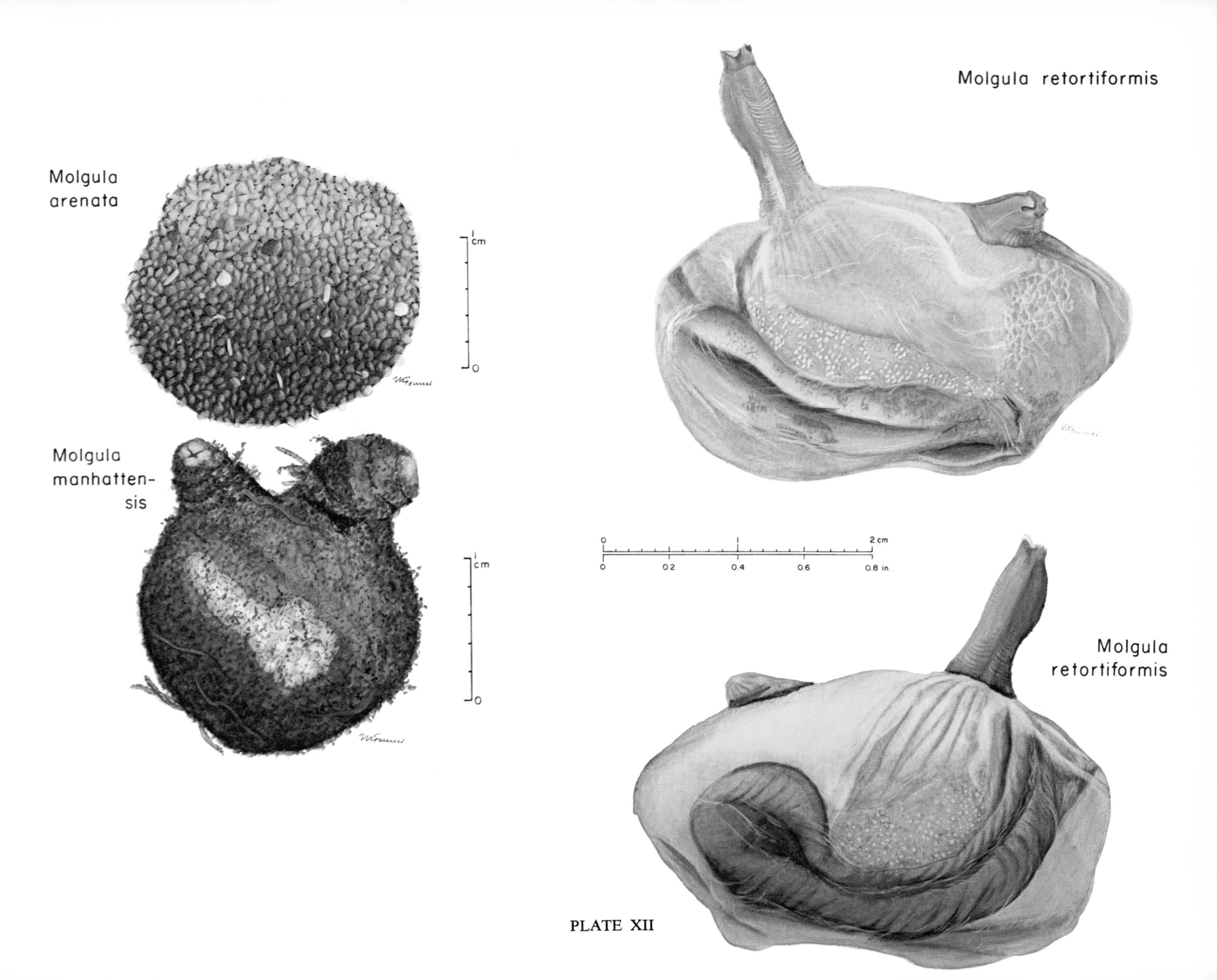
Molgula arenata
cm
0
Molgula manhatten-sis
cm
0
Molgula retortiformis
0
2 cm
0
0.2
0.4
0.6
0.8 in.
Molgula retortiformis

PLATE XII

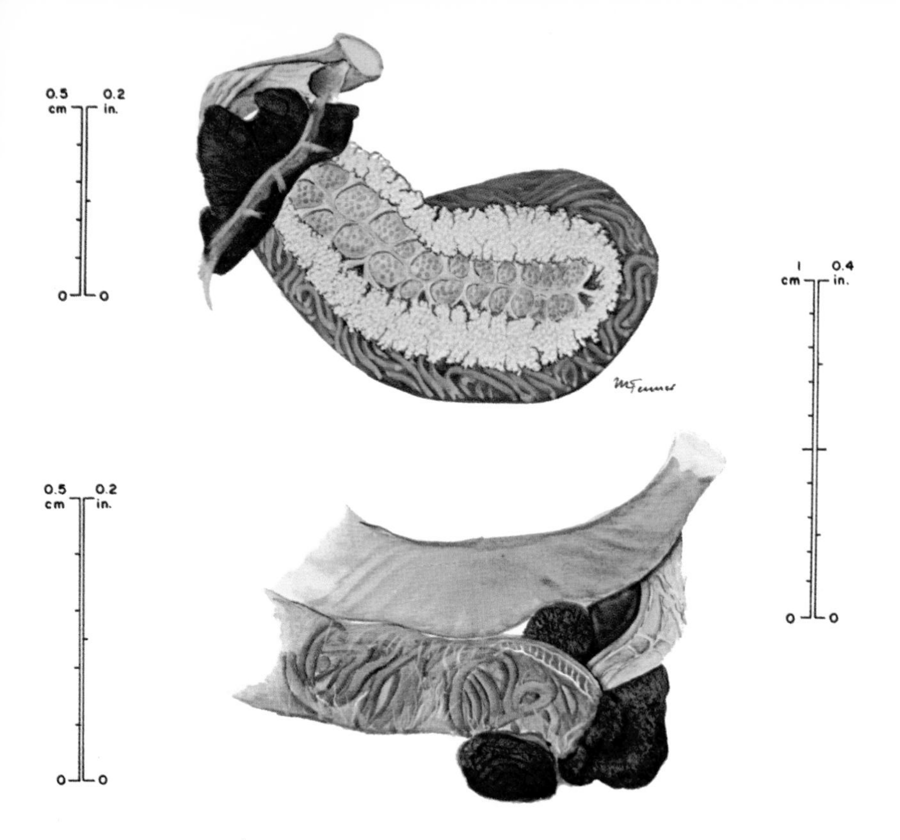

Bostrichobranchus pilularis

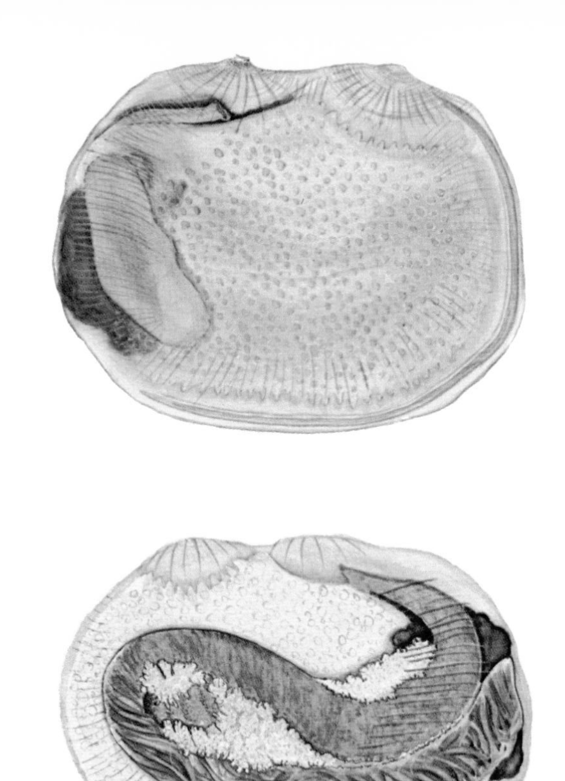

Bostrichobranchus pilularis

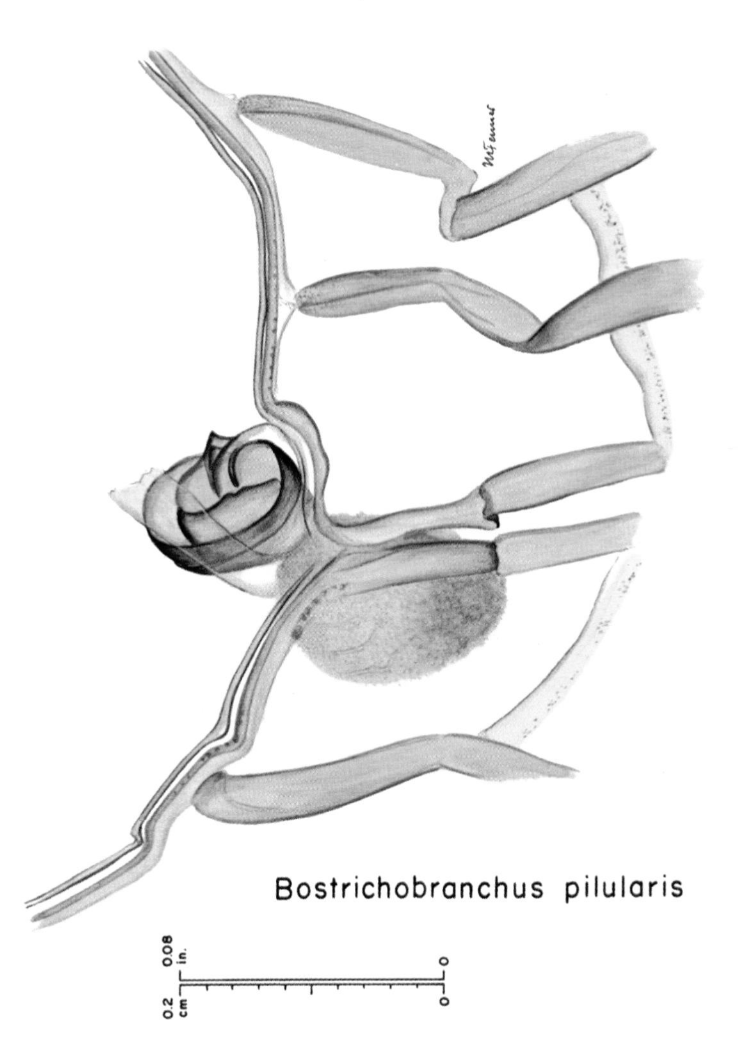

Bostrichobranchus pilularis

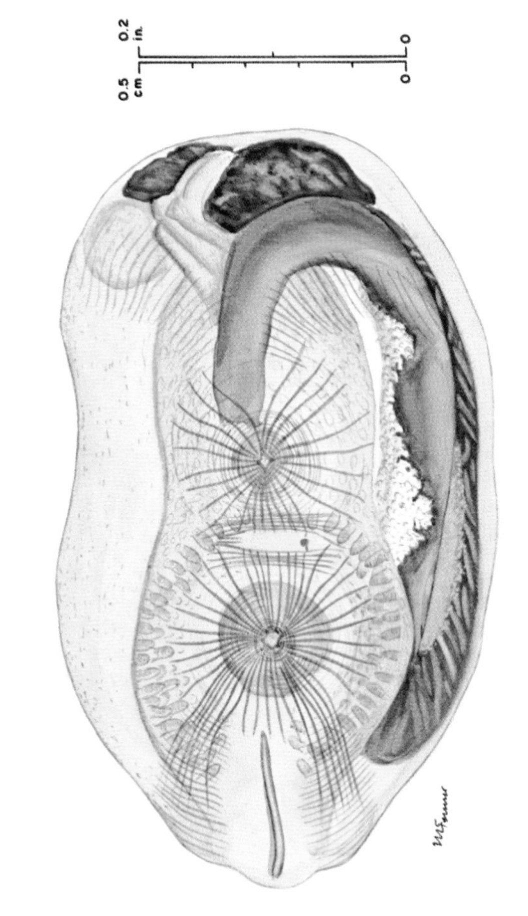

Bostrichobranchus pilularis

PLATE XIII

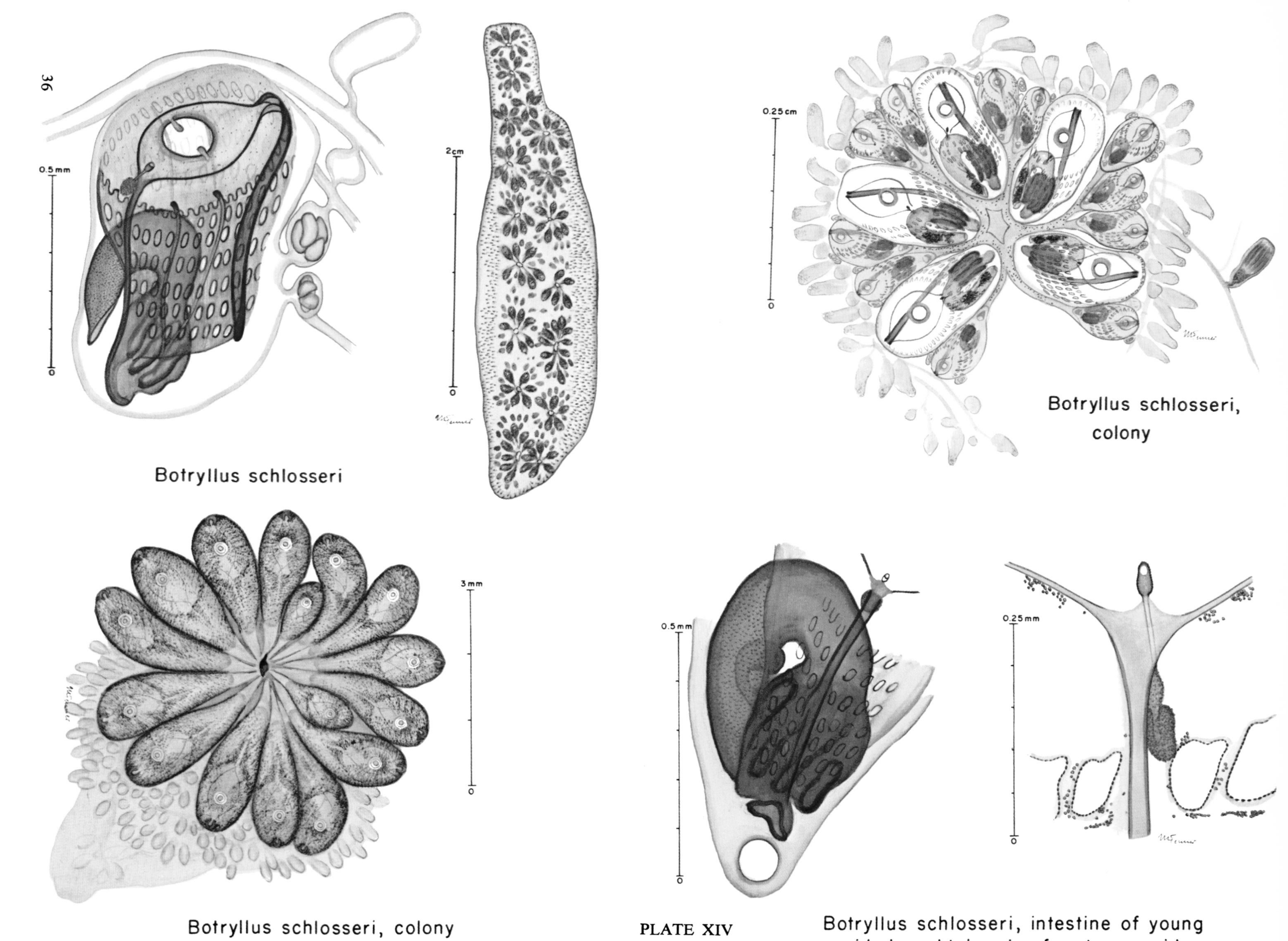

Botryllus schlosseri

Botryllus schlosseri, colony

Botryllus schlosseri, colony

PLATE XIV

Botryllus schlosseri, intestine of young zooid, dorsal tubercle of mature zooid

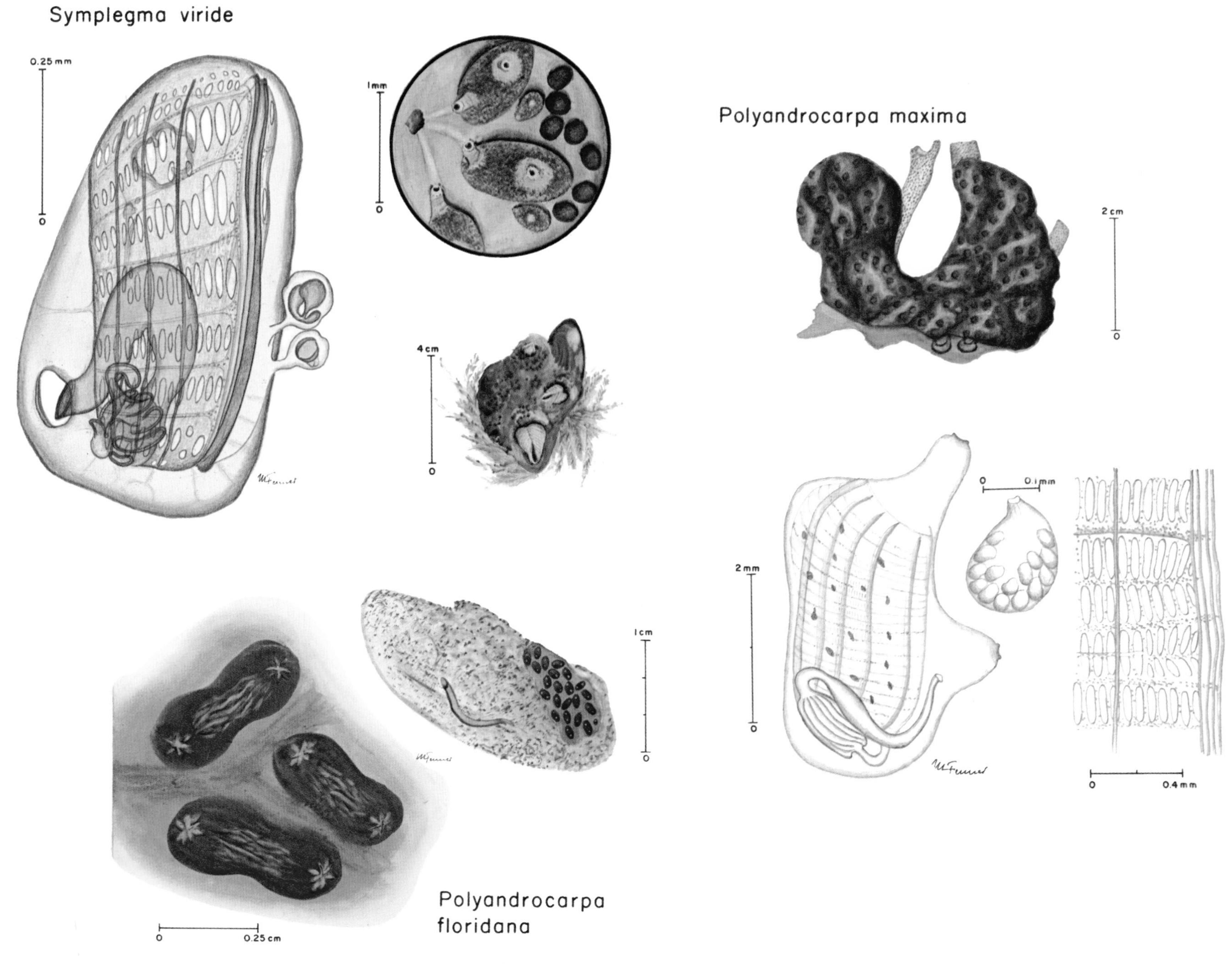
Symplegma viride
0.25 mm
0
1mm
0
4 cm
0
Polyandrocarpa maxima
2 cm
0
0
0.1 mm
0
0.4 mm
2 mm
0
1 cm
0
0
0.25 cm
Polyandrocarpa
floridana

PLATE XV

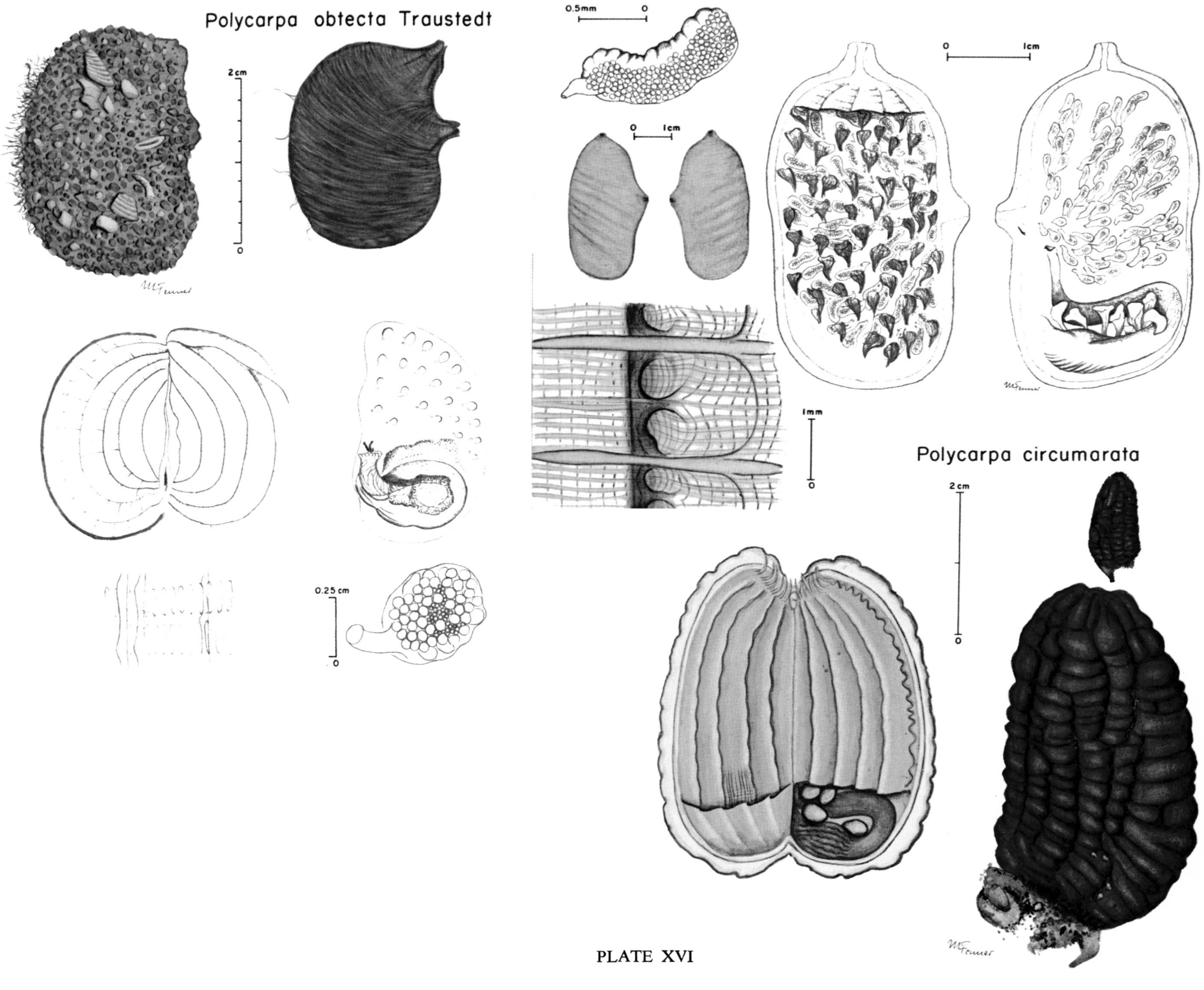

Polycarpa obtecta Traustedt
2 cm
0
0.5mm
0
0
1cm
0
1cm
1mm
0
0.25 cm
0
Polycarpa circumarata
2 cm
0

PLATE XVI

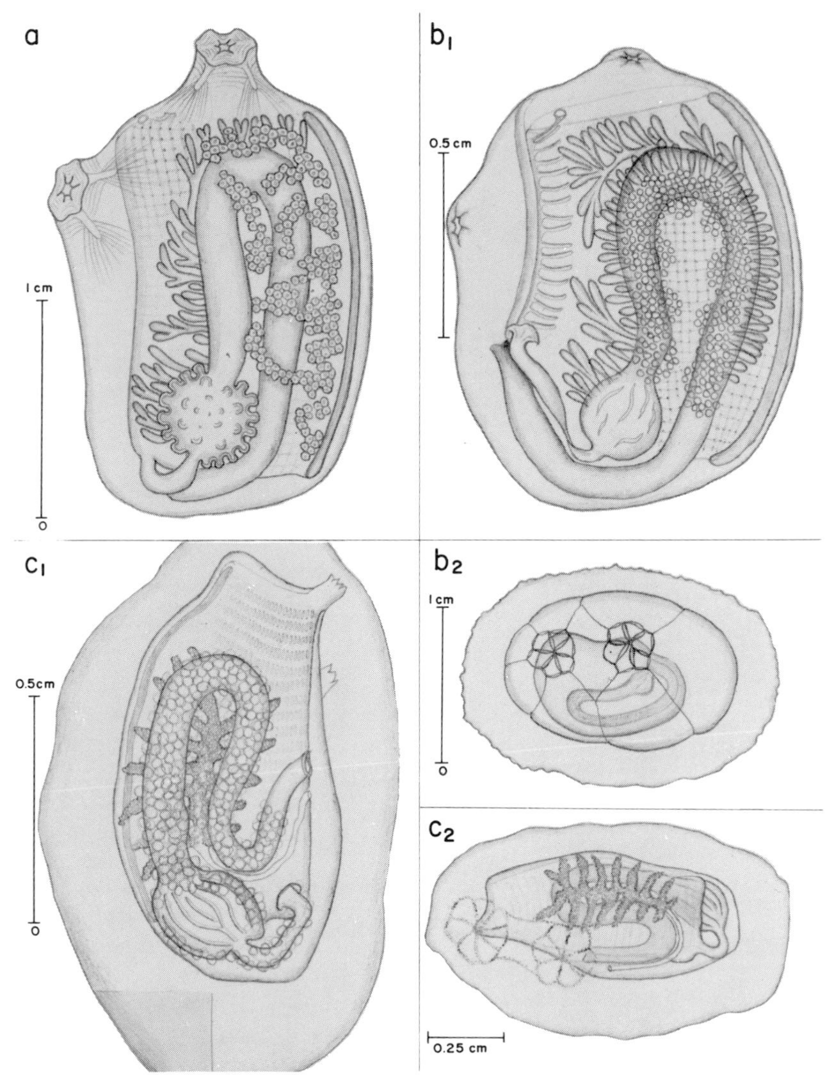

Fig. 14. (a) *Corella borealis*; (b_1) *Chelyosoma macleayanum;* (b_2) same, outside, left; (c_1) *Rhodosoma wigleii*, sp. nov.; (c_2) same, outside, right.

The two other subfamilies form buds and grow into extensive encrusting colonies. The first is found only in shallow water in warmer coastal waters and is the subfamily *Polyzoinae*. Off the Florida west coast are found tough, encrusting, dark red or brown masses within which are enclosed a layer of zooids each of which is very like one of the solitary species, growing singly and much larger in other areas. These are species like *Polyandrocarpa maxima*. Often mixed with and overgrown by this is another quite different colonial species, *Symplegma viride*. These form distinctive colonies by buds at the bases of parent zooids (cf. Plate XV).

The main stem of the *Styelinae* consists of solitary species living close to shore, like *Styela partita*, (Fig. 16) from Maine to Florida, or *Styela plicata* (Fig. 17) in the south. But there are species which prefer colder, deeper water, like *Polycarpa fibrosa*, and these may have persisted from the continental split (Fig. 18). Possible descendants in the south are *Polycarpa circumarata* and *P. obtecta* (Plate XVI), which are found in shallow water. There are many other kinds of Styelids which are loosely attached to sand in deeper water, like *Cnemidocarpa mollis* (Fig. 19). Other divergent species like *Dendrodoa carnea* (Fig. 20) have adapted a tight, flattened attachment to glacial pebbles in moving water currents, or to deep sea habitats where the food particles sink slowly from above. Finally, there are some deep sea Styelid-related species which are dredged

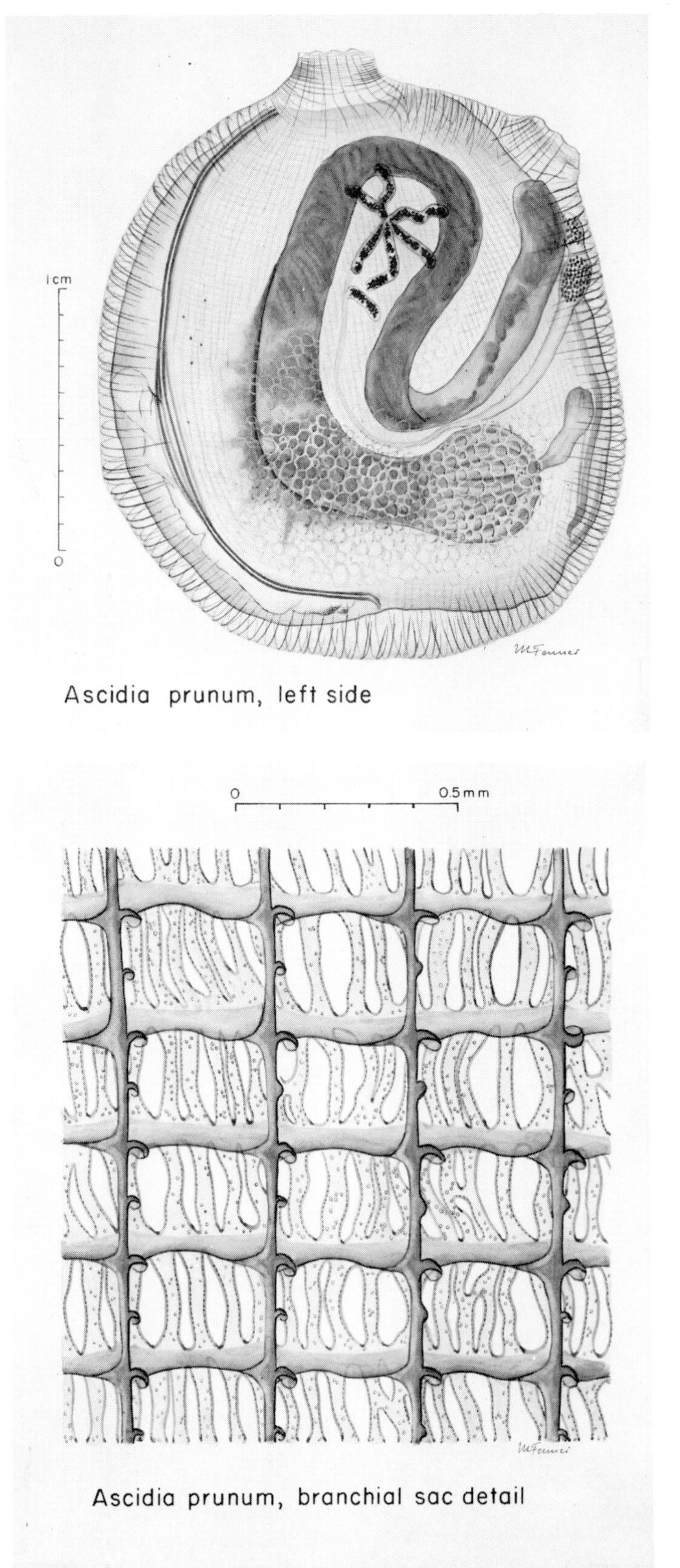

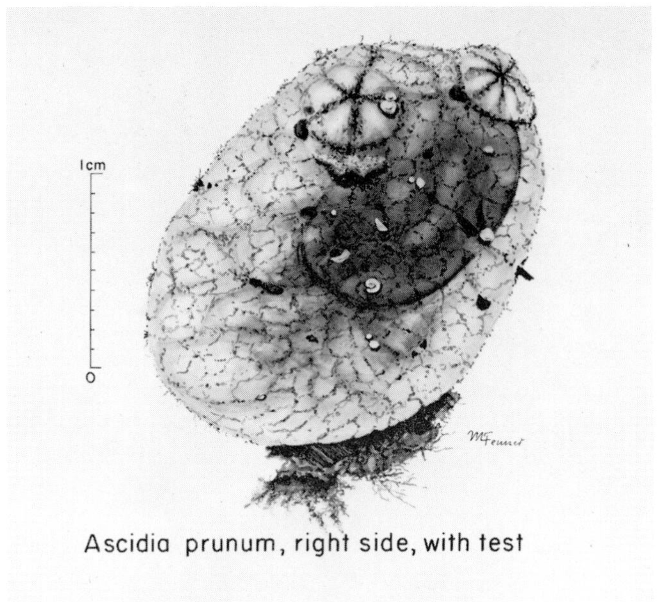

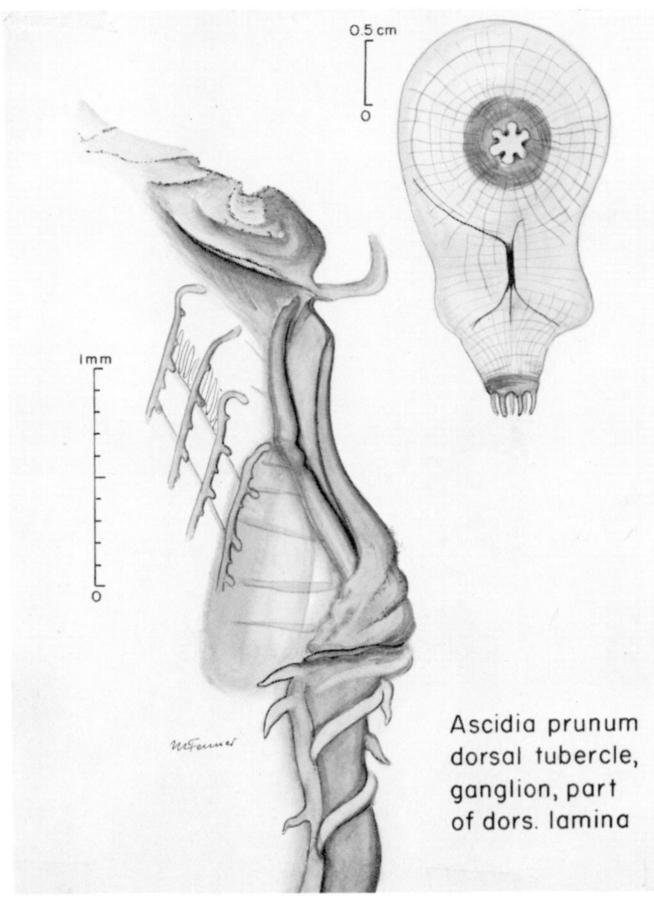

Fig. 15. *Ascidia prunum.*

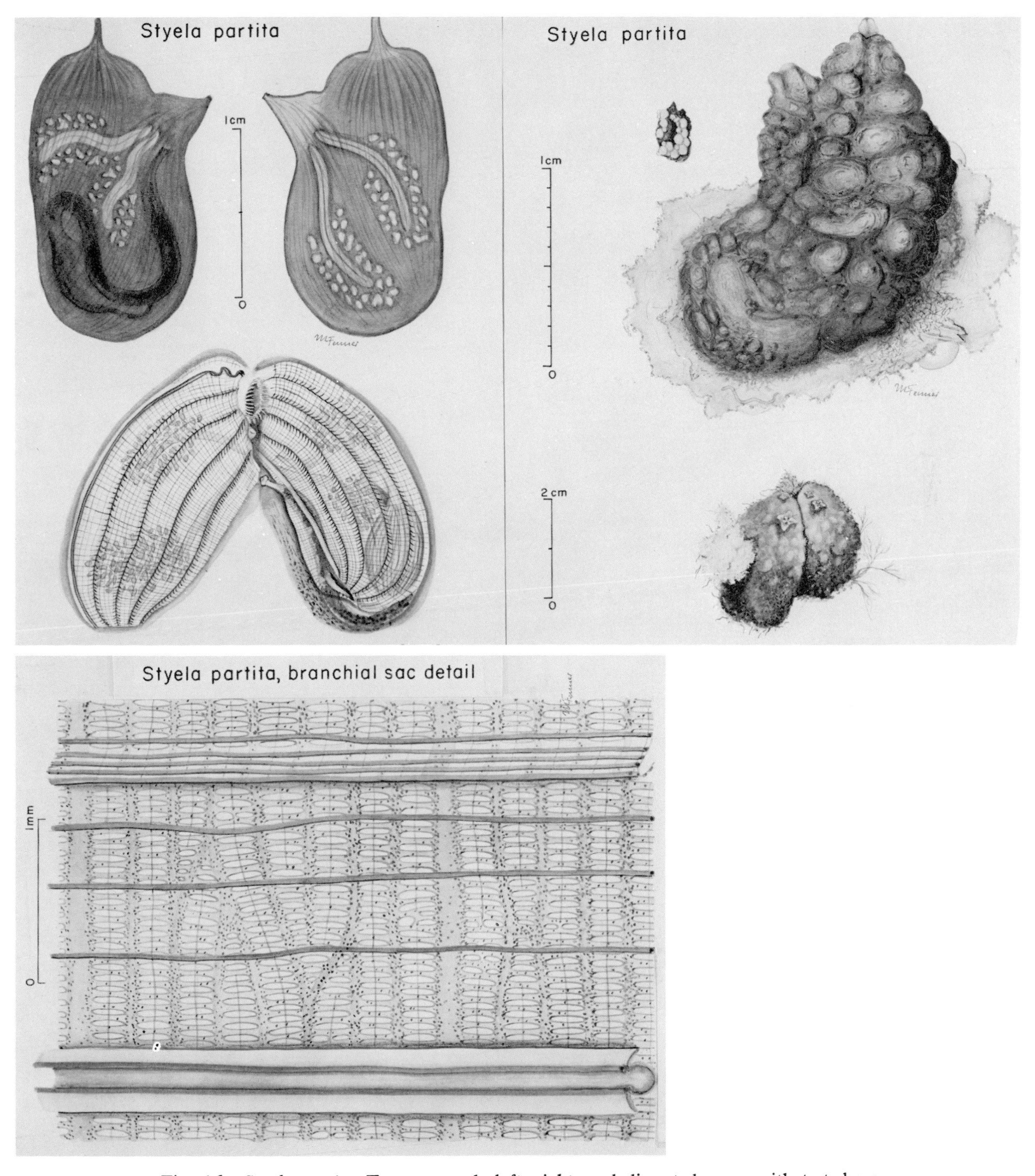

Fig. 16. *Styela partita*. Test removed: left, right, and dissected open; with test; branchial sac detail.

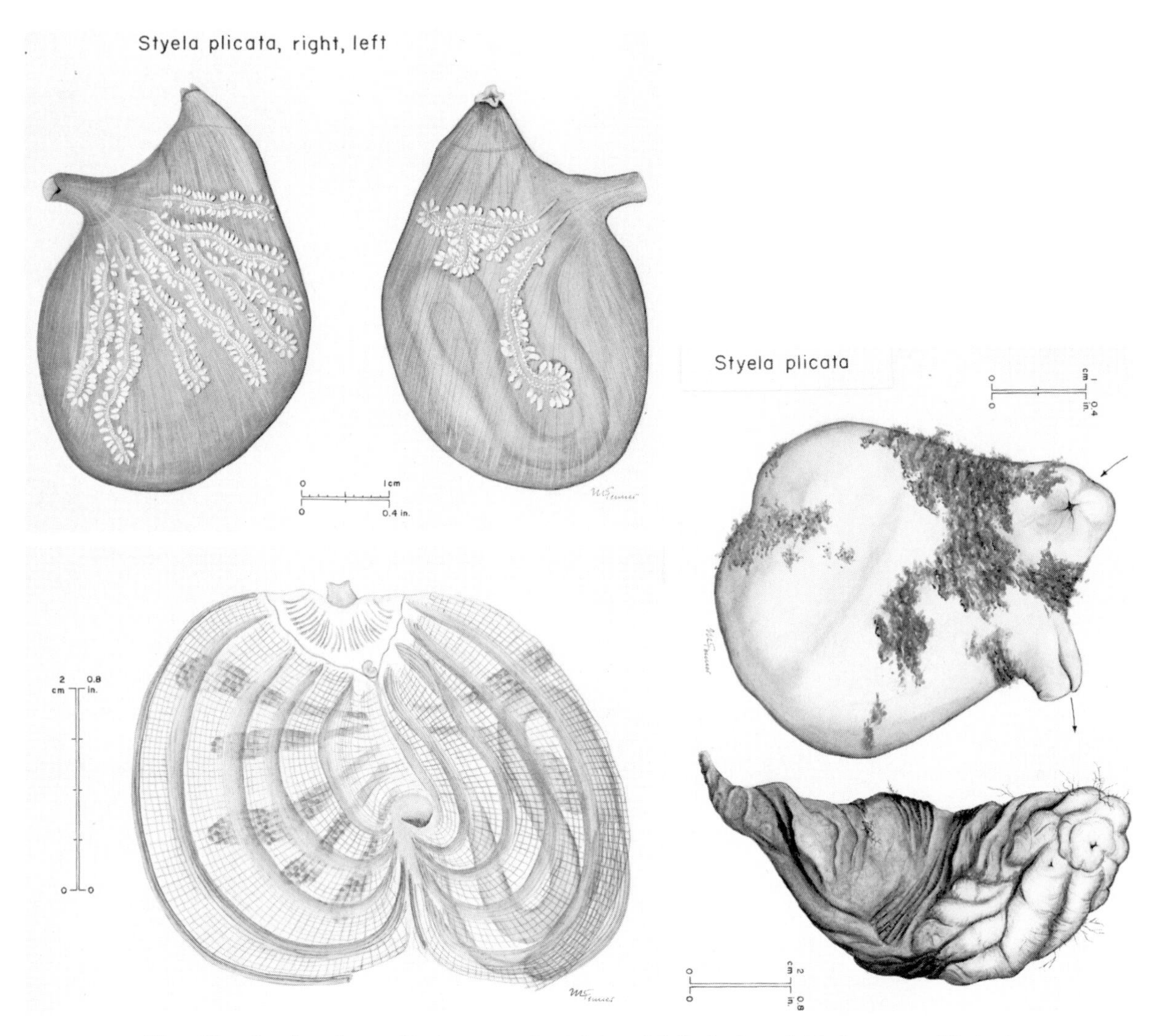

Fig. 17. *Styela plicata.* Test removed: right and left; inside, folded open; with test.

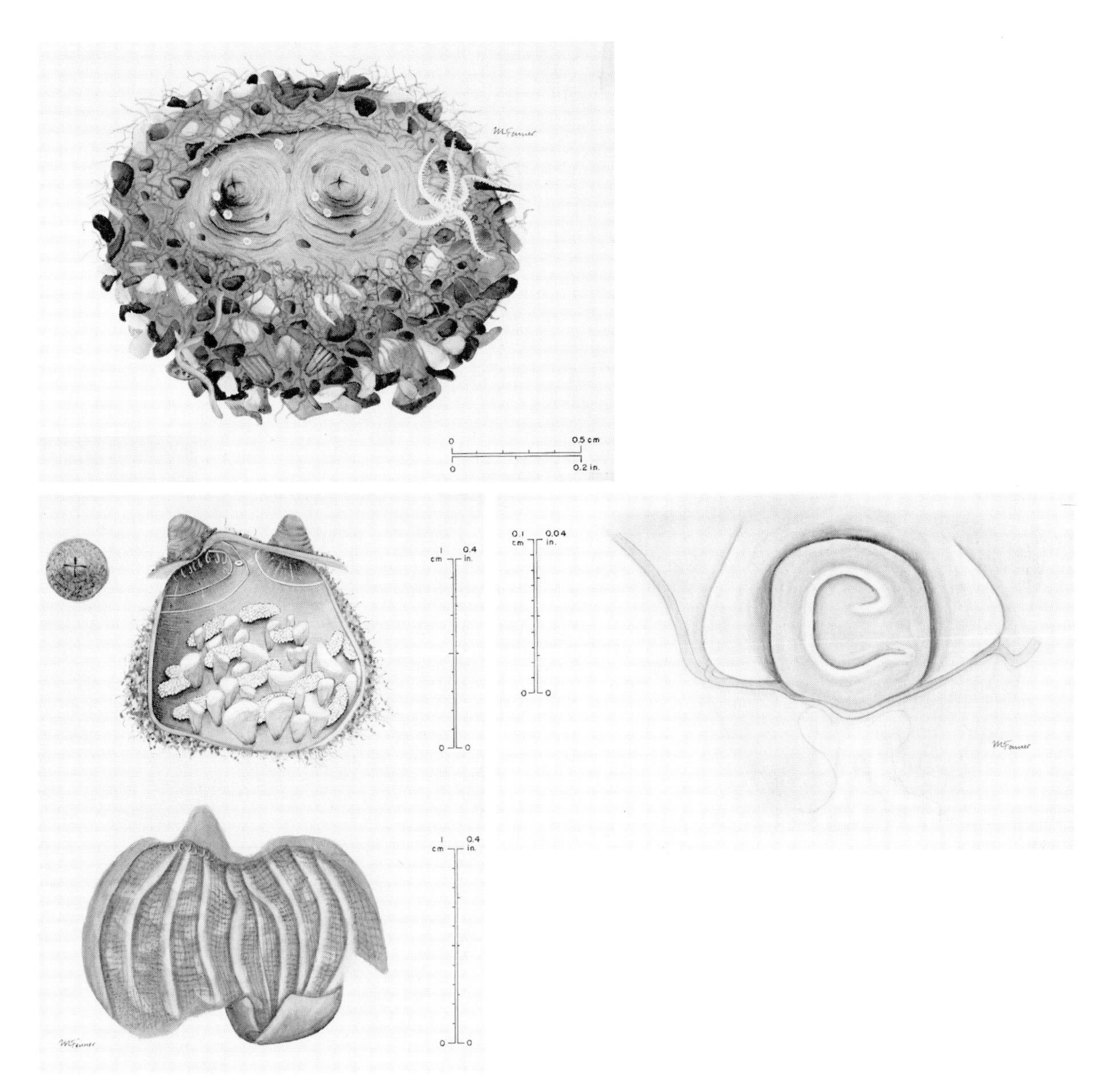

Fig. 18. *Polycarpa fibrosa.* As found; test and gonads, branchial sac opened; dorsal tubercle.

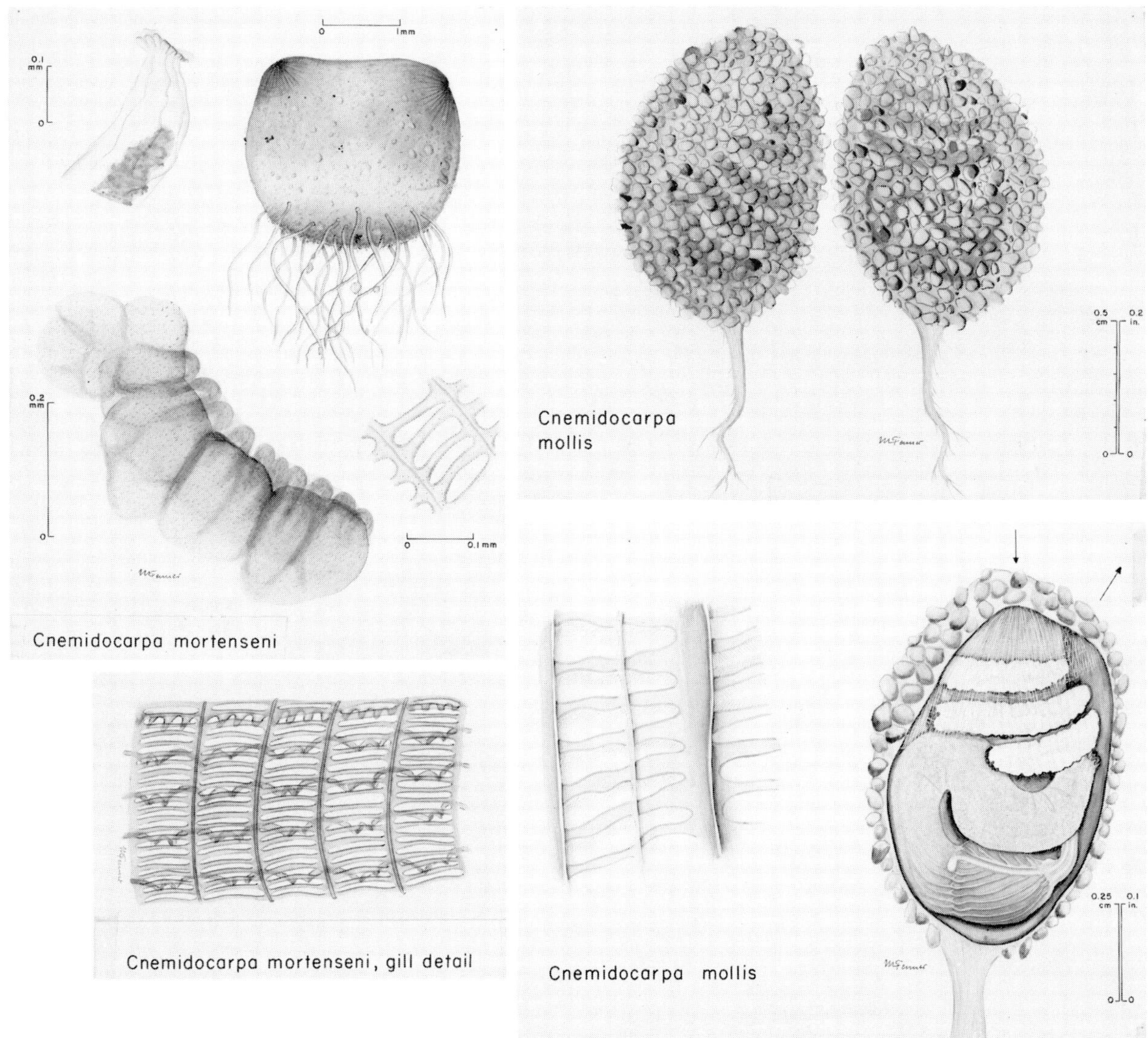

Fig. 19. *Cnemidocarpa mortenseni*. As found, plus various details, including gills. *C. mollis*. As found (right and left), test removed, detail of stigmata.

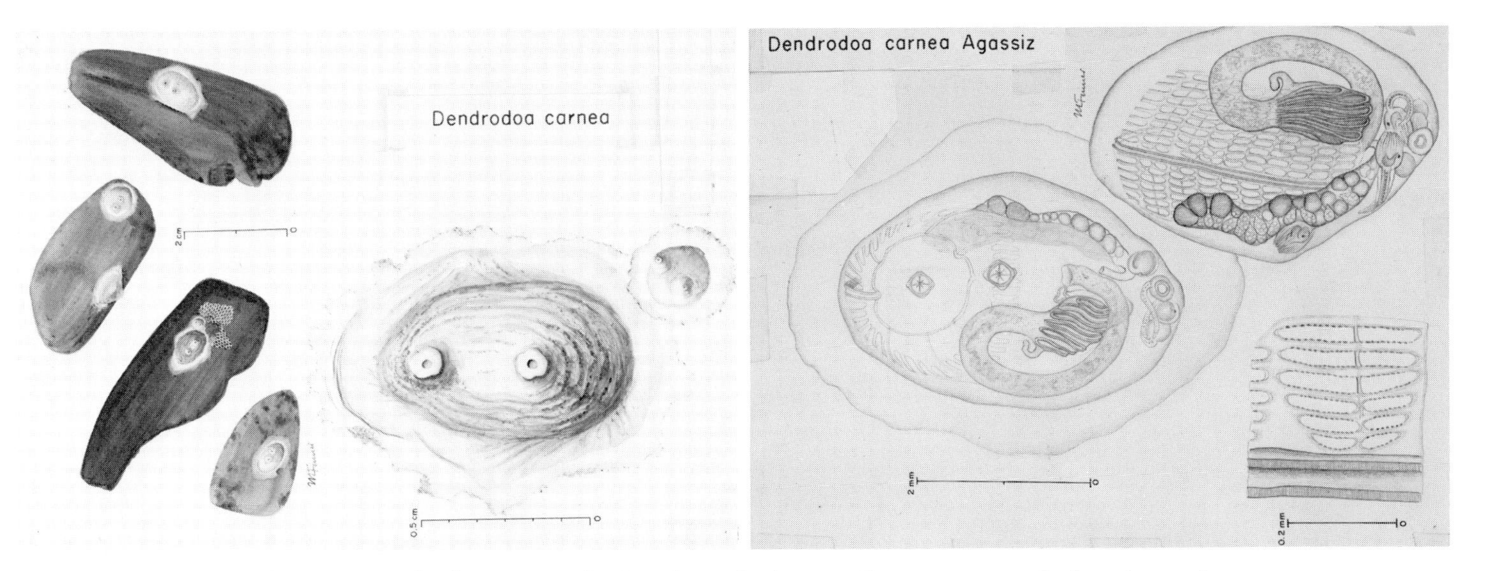

Fig. 20. *Dendrodoa carnea.* As found attached to small stones; external view; internal views, top and bottom; gill detail.

from many deep trenches, such as the *Dicarpa simplex* (Fig. 21), in which the branchial sac has become only a net.

The typical Ascidian tadpole larva, with two sense organs, the eye spot, and otolith, is formed from eggs in the *Enterogona*, like *Ascidia*. Among the *Pleurogona* such a primitive larva is formed only in a few of the *Pyuridae*. In each of the other two major families the tadpoles are somewhat modified or suppressed altogether. It is in the *Molgulidae* and *Styelidae* that the most extreme modifications of the straining clefts or stigmata in the wall of the branchial sac have occurred. Illustrations show some of the different arrangements which have evolved in the large subfamily *Styelinae*.

The more recent experiments in evolutionary change, as in the *Pleurogona*, are far removed from those shown in *Aplousobranchia* and *Phlebobranchia*, which spread over the more restricted continental shelves of Paleozoic times. So it seems probable that the Ascidian ancestors of the earliest *Chordata* were either members of the *Aplousobranchia*, like *Ciona*, or among the early *Phlebobranchia*, like *Ascidia* or *Corella*.

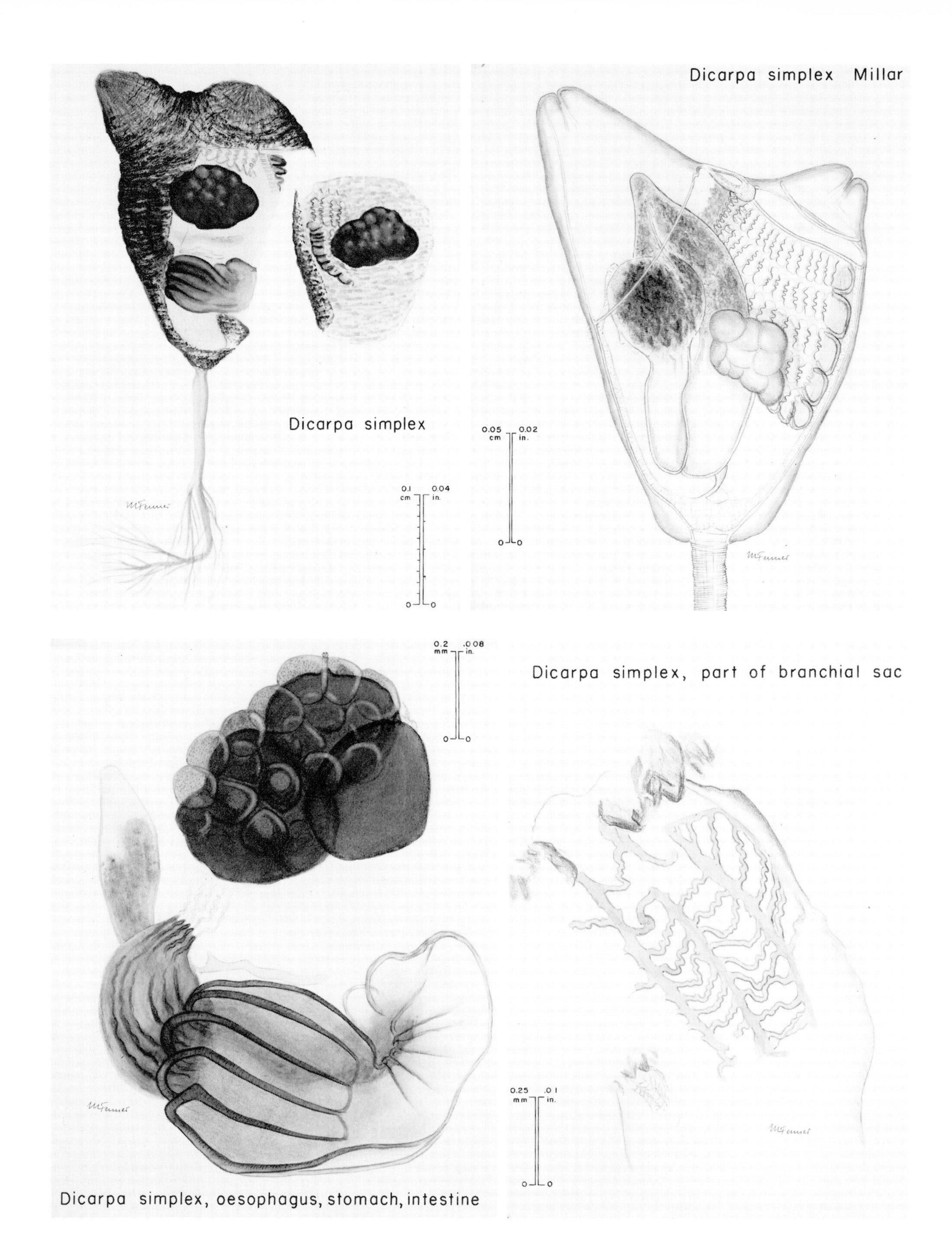

Fig. 21. *Dicarpa simplex.* Test cut open; internal view of zooid; esophagus, stomach, intestine; branchial sac detail.

V. Modifications of the Branchial Sac of Evolutionary Significance

In the previous chapter all of the different families of *Ascidiacea* have been reviewed. It has become evident that even in this small class of *Chordata* a number of different lines of evolutionary change are represented. Each family is a series of species which meet the basic needs for food and oxygen in the marine environment in different ways. Even with small environmental differences various devices appear within each family which increase the effectiveness of the branchial sac in meeting the animal's needs. These structural and physiological differences appear to be genetically determined, and those species which possess them under the matching environment have an advantage shown by wide dispersion and large numbers of offspring. This differential survival in subsequent descendants illustrates determining factors for evolutionary change in Ascidian families. It suggests that this has been determined by a number of different lines of differentiation in the branchial sac.

A brief review of some of these advantageous experiments in different Ascidian families seems desirable before a complete survey of all the species of *Ascidiacea* on the Atlantic continental shelf is undertaken. The structure of the branchial sac in twelve different species in ten different families or subfamily groups is described briefly. Included are figures which will make a little more obvious the family and genus differences already illustrated in plates of family representatives in the previous chapter. It has been emphasized, beginning with the structural review in the first chapter (cf. Figs. 1, 2a, & 2b), that in all Ascidians it is the branchial sac which is the source of food and oxygen for the zooid. This organ maintains a ciliary controlled current drawn in at the oral siphon and out through the atrial opening. In addition, it maintains the rotating mucus rope which picks up the tiny particles, mostly phytoplankton, carried in the moving sea water. Counts of these nutrient particles show a wide range in numbers of particles in different oceans of the world, averaging better than 20 per liter off Cape Cod.

In his volume *Biology of Suspension Feeding* (1966), C. R. Jorgensen sums it up as follows (quoted from the author's article in *Biological Reviews* 1956, 30: 391–454, which is a little more compact):

> Most sponges, lamellibranchs and ascidians filter 13–16 liters of H_2O for each ml. O_2 taken in. Suspension feeders which filter 15 liters for O_2 uptake can cover their food requirement for optimal growth from that amount of sea water. Thus 0.15 to 0.20 mg. of utilizable organic matter is an average amount from 16.6 liters of H_2O. It appears that the amounts of phytoplankton in coastal waters is ample. At great depths new figures are needed. Here the problem is to get sufficient oxygen. Here animals must filter 100 liters of H_2O to get 1 ml. of O_2—or seven times more than in coastal waters.

The depth appears to be the environmental condition which is of greatest importance in determining the amount of tiny food particles in the sea water as well as the oxygen in solution. Various devices in the different families increase the effectiveness of the zooids in the environment where they are found, but the organ of most importance is the branchial sac. It will be helpful to compare the families and subfamilies by examining the branchial sac in a number of well-known species. This is done in Table II and in the accompanying figures, each of which is a composite (Figs. 22, 23a–f, & 23g).

Each of the species listed in Table II and illustrated in the figures emphasizes some distinctive characteristic of the branchial sac. In Figures 22 and 23 it will be useful to note in the sac the increase in size, the increased number of slits or stigmata, the marked increase in the longitudinal and transverse blood vessels in the sac wall, and the longitudinal folds which determine the number of bundles of stigmata. Another development which was of evolutionary importance was that the surface of the sac has been increased in area by elongation and by fluting of the surface in waves. Finally, there have appeared several patterns of extension and spiralization of the stigmata which result in longer slits, or even tent-like spiral elevations of the stigmata themselves into what are called infundibula. Some of these modifications of the surface of the branchial sac are

Table II. Comparison of Branchial Sacs in Representatives of Ascidian Families

Figure	Family	Species	Median depth at which it grows	Distinctive character
22a	*Holozoinae*	*Distaplia bermudensis*	5 m	short sac, 4 rows stigmata
b	*Polyclinidae*	*Aplidium glabrum*	50 m	long sac, 10 rows stigmata
c	*Cionidae*	*Ciona intestinalis*	25 m	no folds, longitudinal vessels present
d	*Perophoridae*	*Perophora viridis*	5 m	4 rows longitudinal vessels present
e	*Ascidiidae*	*Ascidia callosa*	50 m	no folds, many stigmata
f	*Bolteniinae*	*Boltenia ovifera*	100 m	all stigmata placed transversely
23a	*Molgulinae*	*Molgula citrina*	5 m	7 rows, spiral stigmata
b	*Eugyrinae*	*Bostrichobranchus pilularis*	15 m	spiral stigmata in elevated infundibula
c	*Polyzoinae*	*Polyandrocarpa maxima*	5 m	4 folds, simplified structure
d	*Styelinae*	*Cnemidocarpa mollis*	30 m	4 folds, 10 stigmata between vessels
e	*Styelinae*	*Stylea plicata*	10 m	4 folds, longitudinal vessels few
f	*Styelinae*	*Dicarpa simplex*	1800 m	simplied 4 widened stigmata
g	*Molgulinae*	*Molgula citrina*	5 m	elongated cilia in stigmata

characteristic of every Ascidian species, but they will be little understood except by examining specimens with the assistance of a series of diagrams like those in Figures 22 and 23. The basic structures can be seen by reference to the generalized structural diagrams in Figure 1.

In Figure 22a, *Distaplia bermudensis*, the sac is wide and rather stiff, with only four rows of stigmata, in size about 2 × 1 mm, and having a tubelike incurrent siphon. The water passes through the fine stigmata into the narrow atrial cavity surrounding and out by the funnel-shaped excurrent siphon. There is a wide groove at the lower (right) edge of the sac, the endostyle, which secretes mucus into the cavity of the sac. The tiny particles carried by the entering sea water are entangled in this mucus mass, which is fed into the lower tube or esophagus. This mucus rope is the continuous source of food. It is shut off only when a sudden muscular contraction shuts the whole sac and closes the entering siphon (cf. Fig. 1). This may be the earliest type of branchial sac, although it is possible that it has arisen by shortening that in the next species.

The branchial sac of *Distaplia bermudensis* is the simple water-collecting apparatus such as is found in all animals which are suspension feeders. It is small, flexible, and bears only four rows of small stigmata. This seems to have been the kind of branchial sac found in the earliest Ascidian species. Indeed, it may be that the very earliest Ascidian species was a bright-colored mass of Distaplia-like zooids attached to stones or seaweed in water about 5 to 10 meters deep. Another similar possibility is suggested by the immense rope-like floating colonies in antarctic seas of *Distaplia cylindrica* (cf. Kott, *Antarctic Ascidiacea*, p. 30).

There is another family of *Ascidiacea* which is structurally quite simple. It is the family *Polyclinidae*, and it includes such species as *Aplidium glabrum* (Fig. 22b) and *A. pallidum*. These species are found in cold seas all over the world, usually in a little deeper water than are the *Holozoinae*. They are colonial Ascidians, often forming massive bright-colored masses containing elongated zooids. These elongated zooids possess a long branchial sac with sometimes as many as sixteen rows of stigmata with an abdomen and post abdomen. The long collecting sac emptying into a common cloacal chamber is an even more efficient apparatus for sifting large amounts of sea water containing food particles and dissolved oxygen than the shorter tubes of *Distaplia*. This family, like the previous one cited, contains only colonial species. These long branchial sacs of *Aplidium* may have been the earliest forms which gave rise to the shorter sacs of the *Distaplia*.

Solitary species of Ascidians show a wide diversity in the structure of the branchial sac, developing it into an efficient collecting organ in a number of divergent directions. Many of these evolutionary lines are shown by the examples listed in Table II, and also shown in Figures 22 and 23. As a result of these many different experiments, the *Ascidiacea* have become successful enough to persist as benthic marine animals on all continental platforms through more than a hundred million years. The major special features of the more widespread families are discussed briefly below.

In Figure 22c, *Ciona* has greatly enlarged the branchial sac, with a large number of irregular stigmata, but there are no branchial folds. In contrast, *Perophora* (Fig. 22d), like *Distaplia*, retains the small number of stigmata arranged in rows, and in addition has longitudinal blood vessels in the sac wall. The *Ascidiidae* (Fig. 22e) has become a widespread family with many

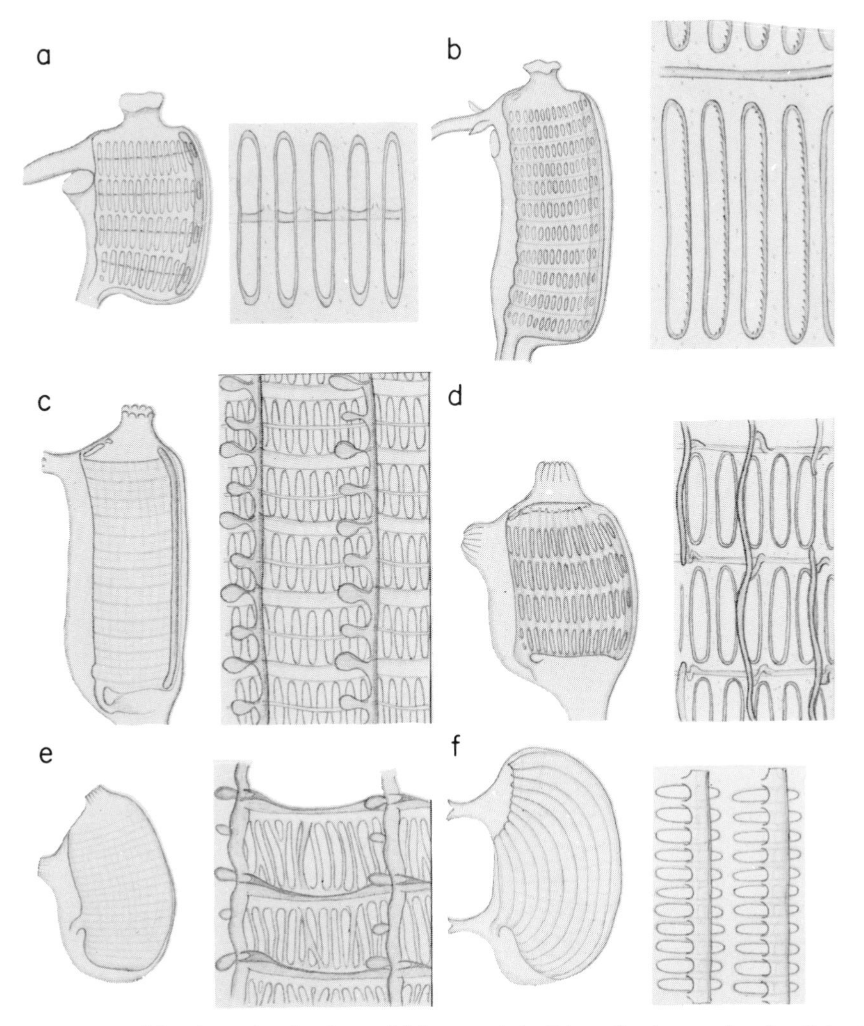

Fig. 22a–f. Modifications in the branchial sac. (a) *Distaplia bermudensis*; (b) *Aplidium glabrum*; (c) *Ciona intestinalis*; (d) *Perophora viridis*; (e) *Ascidia callosa*; (f) *Boltenia ovifera*.

species spread over the continental shelf in northern areas in deeper water and in southern latitudes close to shore. Here the number of stigmata is greatly increased, and they are arranged in regular rows separated by longitudinal blood vessels and fluting of the sac surface, but again no folds.

Beginning with the *Pyuridae* in the subfamily *Bolteniinae*, we come to the most advanced of the Ascidian families which have made structural advances in the branchial sac of sufficient value to give them a widespread competitive position on continental shelves all over the world. In the deeper water the Gulf of Maine *Boltenia ovifera* grows attached by a long stalk. Its branchial sac (Fig. 22f) develops nine or ten deep longitudinal folds within which are a greatly multiplied number of fine stigmata running transversely to the axis and the longitudinal blood vessels. In addition to a great multiplication in the number of stigmata, their transverse position suggests the enlargement by spiralization of the branchial stigmata shown by the *Molgulidae*.

Passing on to the subfamily *Molgulinae* we see the branchial sac with spiralized stigmata in every member species. Representatives live in shallow or slightly deeper water in every habitable region of the world. They are the most widely distributed of Ascidian genera. *Molgula citrina* (Figs. 23a & 23g) is a shallow water species found in sheltered patches along the northern Atlantic continental shelf, where there is effective water movement. It forms a large actively swimming tadpole larva which is frequently found in summer collections. Although active, this tadpole possesses only one of the two original sense organs, the otolith. Its branchial sac has seven folds, and the typical spiral stigmata. In the

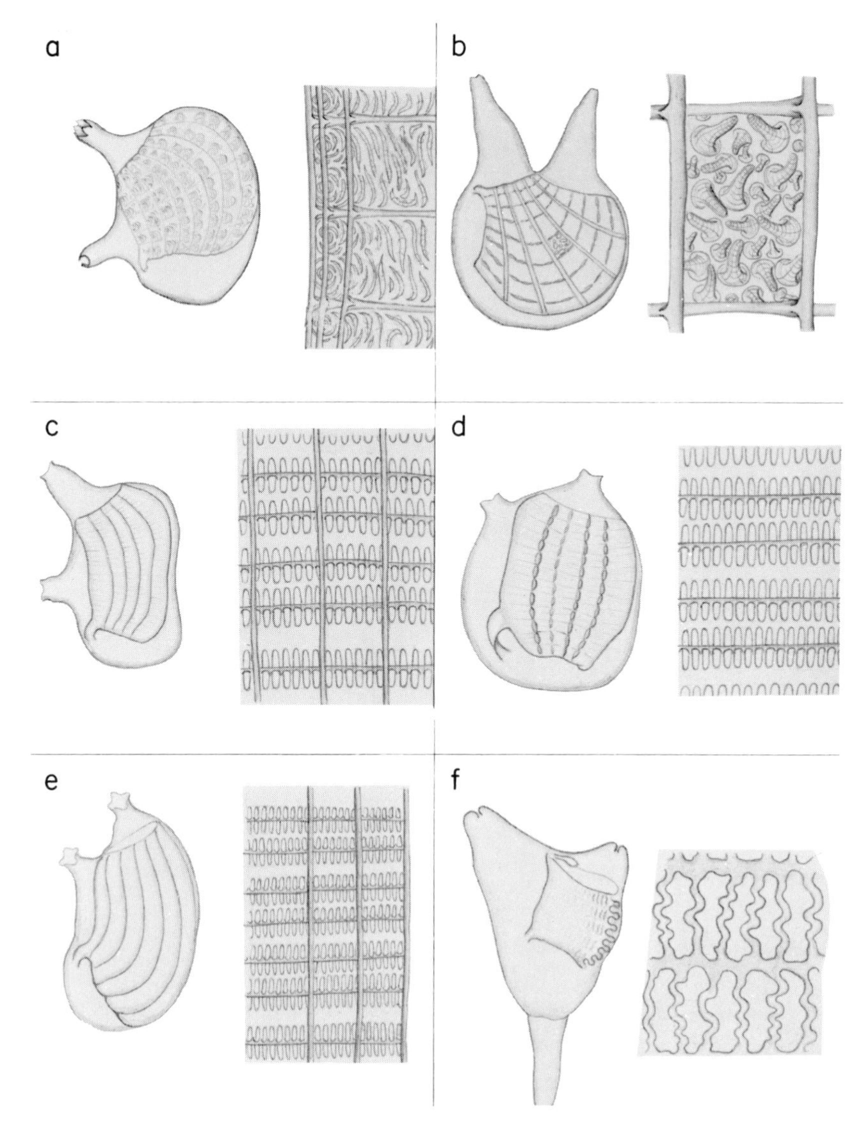

Fig. 23a–f. Modifications in the branchial sac. (a) *Molgula citrina*, bent slits; (b) *Bostrichobranchus pilularis*, spiralization; (c) *Polyandrocarpa maxima*; (d) *Cnemidocarpa mollis*; (e) *Styela plicata*; (f) *Dicarpa simplex*, net, deep water.

same family but listed in another subfamily, *Eugyrinae*, is a species *Bostrichobranchus pilularis* (Fig. 23b) distributed at intervals off the whole eastern coast, which shows the spiralization of branchial sac stigmata carried to its extreme. The slits become inverted corkscrews elevated into a succession of more and more tightly coiled infundibula, which give the sac surface a bizarre appearance. This great extension of the surface of the branchial sac allows still greater food and oxygen interchange in all the species of *Eugyrinae*. This is the adaptation which aids this species to live in muddy shallow spots all along the Atlantic continental shelf from off Maine to Padre Island off Southern Texas.

A different series of special structures can be seen in the branchial sacs of certain species of *Styelidae*. In this family are found a few colonial species, like *Polyandrocarpa maxima*, in which the small zooids show smooth branchial sacs included in the much larger solitary species. This fact seems to suggest that the colonial species may have been derived from the solitary species, like *Polycarpa obtecta* (cf. Fig. 23c). They both live quite close to shore, especially off the Florida west coast, and also Louisiana and Texas in the Gulf of Mexico.

The three species yet to be mentioned are all members of the *Styelinae* and illustrate the great diversity of

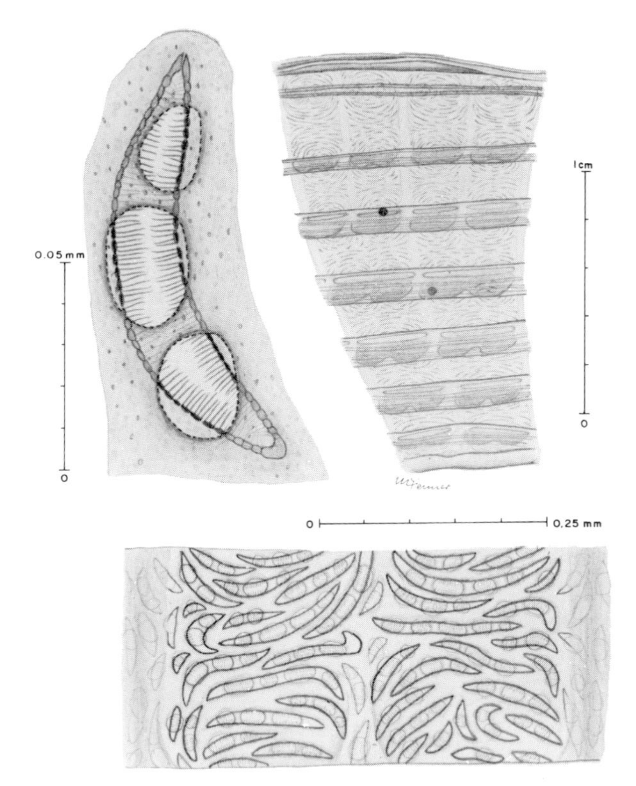

Fig. 23g. Modifications in the branchial sac. *Molgula citrina*, branchial slits at increased magnification.

species within this subfamily. There are representatives of two of the commoner in-shore species, *Cnemidocarpa mollis* and *Styela plicata*. Branchial sac drawings are shown in Figure 23d & e. Both of these animals show four large folds in the sac, with very large numbers of wide straight stigmata. *Styela plicata* is the large southern species living close to shore, tightly attached to stones or shells, and with a powerful food current through its siphons. *Cnemidocarpa mollis* lies on the sand in deeper water. It is more numerous than *Styela* and maintains a visible food current because it is present in large numbers on favorable sand. The Styelids have developed a large, smooth branchial sac which is effective where there is ample food. Thus the large sac is effective in deep water with *Polycarpa fibrosa* and in rough water on northern sand in *Pelonaia corrugata*.

Finally, the *Styelinae* include species of *Dendrodoa* which have become adapted to life while rigidly attached on one side to stones placed in a rapidly flowing current. The flattened existence has resulted in the loss of branchial folds and one of the original gonads. But *Dendrodoa carnea* has held on to its wide, smooth branchial sac, and with its simplified structure, it remains effective off Block Island, in the Gulf of Maine and farther north.

Another reasonably effective Ascidian related to *Styela* is found by dredging in deep water either on the continental slope, or even off the Atlantic shelf at depths of 1500 meters or more. This species is very small—10 mm—and is attached by a stalk of about the same length in the dark abyss. This tiny Ascidian species is called *Dicarpa simplex*, and it has been dredged up from off North Carolina and other deep trenches, as well as one hundred miles off shore in the latitude of the Hudson and the Baltimore canyons. An especially fine specimen dredged by the Duke University Oceanographic Unit is shown in Figure 21, and an outline sketch in Figure 23f. It is clear that this little 7-mm zooid has very wide open stigmata, giving its branchial sac almost a net-like appearance. It visually emphasizes Jorgenson's statement that such a deep sea animal must maintain a constantly flowing stream of water through its branchial sac net in order to get sufficient food and oxygen to keep alive in the dark at 5,000 feet depth. By maintaining a constant stream of water through its siphons and its net-like branchial sac, this species is able to live under the extreme conditions found close to the bottom of the continental slope. It is of considerable interest that at these depths there is food and oxygen if only the animal can develop physiological mechanisms to utilize them.

This rapid overlook of a dozen or more Ascidian species in as many families emphasizes the view that it was the enlargement and diversification of the branchial sac which made this class important in the evolution of later *Chordata*. The branchial sac, with the possibilities of its varied and successful experiments with special kinds of gill slits, may have led to a tadpole with a functional mouth and gills derived from Ascidian stigmata.

Figures 22 and 23 also show some of the ways in which the branchial sac was modified in different Ascidian families. The most primitive is seen in *Distaplia* and *Aplidium* (Fig. 22a & b) in which the simple sac, long or short, is pierced by many or few rows of the ciliated clefts or stigmata. In *Ciona* and *Perophora* (Fig. 22c & d), the stigmata are blocked in by longitudinal blood vessels and longitudinal folds. The *Ascidia* species show both longitudinal and transverse blocks of stigmata. In *Boltenia* the experiment was tried of forming the stigmata in a transverse rather than the common longitudinal direction in the branchial sac wall (cf. Fig. 22e & f).

In Figure 23 many further experiments are suggested in arrangement as well as form of the stigmata in the branchial sac. These appear in the *Molgulidae* and *Styelidae*, where the stigmata are separated into blocks by folds. They become elongated and spiralized. Finally, in *Bostrichobranchus* the stigmata become elongated and extended into nipple-like elevations (Fig. 23a & b).

In *Polyandrocarpa* and *Cnemidocarpa* the stigmata remain simple, but the zooids are arranged in colony-like aggregations (Fig. 23c & d). Finally, in *Styela plicata*, the stigmata have been greatly multiplied between blocks of blood vessels. In the deep water *Dicarpa simplex* the reverse process, simplification, has occurred, and the stigmata are widened into a miniature net (Fig. 23e & f).

The changes in the branchial sac show a number of the evolutionary sequences in the *Ascidiacea*. The earliest (Palaeozoic?) species (*Enterogona*) show simple sacs, and it is with them that the primitive tadpoles are found. It is these which might have led to neotenic tadpoles as vertebrate ancestors. The more complicated *Pleurogona* are more efficient "sea squirts," but too specialized to be vertebrate ancestors.

Berrill (1936) summarized the interrelationships of the Ascidian families in his paper on tunicate development (Part V, p. 67). The following arrangement agrees with his summary in all essentials.

Relationships of Families of Ascidians

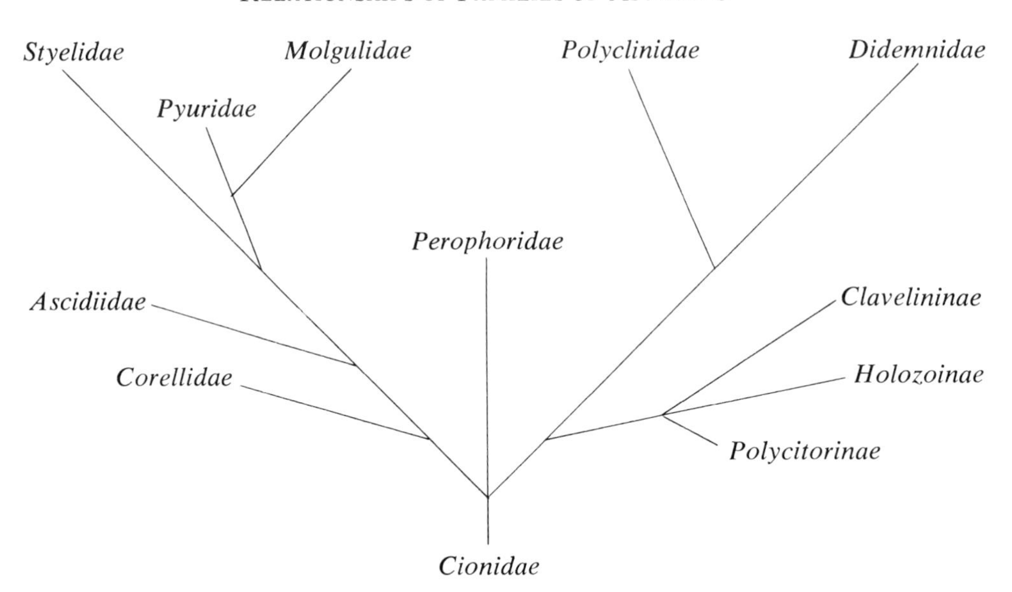

VI. The Species of *Ascidiacea* from Maine to Texas

The distinctive characters of Ascidians in general have been noted in the Preface and Chapter I. Then in Chapters II and III it was shown how Ascidian specimens have been collected on the continental shelf of the United States. It was confirmed that certain well known species are distributed in well-defined areas in northeast or southeast. Continuing in Chapter IV, all the families from which specimens are known on the Atlantic shelf were listed and their distinguishing structural characteristics were briefly summarized. These characters should make it possible to place in the proper family any sea squirt brought in from anywhere on the Atlantic continental shelf of the United States.

The description of Ascidian species of the Atlantic continental shelf is more easily followed if, as in Chapter V, more attention is given to the modifications in the collecting organ of all *Ascidiacea*, the branchial sac. This has changed more than any other part of the simple Ascidian organism in the long period of time during which these relatives of the earliest vertebrates have lived on the continental shelves of the earlier continents of the world.

In this chapter there will be given a brief description of each of the species found on the Atlantic continental shelf in the present new survey. The species is the ultimate taxonomic unit in the biological classification scheme, and each has a distinct structure and distribution. For each of the more widely spread species a brief delimitation of the distribution will be noted, based on our collections and those of certain cooperating naturalists. All of the species found are briefly described under their orders and families and each is described briefly in the text. A complete list of the species of *Ascidiacea* is given in the Appendix.

It is of general interest that the dredging over wide areas of the Atlantic continental shelf off North America confirms the continued presence of most of the Ascidian species described fifty years ago. A few additional undescribed species were collected off Cape Ann, off the Georgia coast, off the west coast of Florida, and off Padre Island, southern Texas. In addition, it has been found that certain well-known species have a wider distribution than has been recorded previously. It should be noted also that the obvious pollution from certain shore situations has not caused the disappearance of a single species, although several in-shore dwelling species have now to be sought some distance outside of previous stated locations.

All *Ascidiacea* are marine, bottom-living or benthic animals, but some are attached upside down under stones or marine protective shelves of various kinds. A number of important species lie on their sides on hard sand with few or no visible attaching filaments. Some so-called solitary zooids, like *Molgula manhattensis*, become associated in great masses, where they may appear like colonies, although each zooid is independent. The distribution of many species is very much wider than casual offshore collecting suggests. *Ascidiacea* are less numerous than *Mollusca*—bivalves or univalves—or *Coelenterata*—hydroids, jelly fish and corals—but every area where these more numerous marine animals live will usually disclose a few delicate sea squirts as well. This includes the tidal regions and the current-bathed expanses a few miles off-shore, as well as the sand or rocky bottoms twenty to fifty miles at sea. Ascidians are well-established inhabitants in certain deep water areas in the Gulf of Maine far east of Cape Cod, also southeast of Cape Hatteras and over the wide shelf off the Georgia coast. There are well-established residents west of Florida at the Tortugas Islands and up along the Florida west coast. Farther west, off Louisiana and Southern Texas, there are bottom-attached Ascidian specimens from tide water out to fifty miles from shore. The newest established United States seashore area at Padre Island, Texas, shows many *Ascidiacea* inside and outside the island, the latter suggesting the West Indian species.

In addition, there are "live bottom areas" several miles off shore at infrequent intervals along the Atlantic coast—usually in sand-covered areas at a turning point in marine currents—which may show a diverse intermingling mixture of a great many Ascidian species found in that particular latitude. In the Gulf of Maine there are outcrops of rock built by ice deposits which form widely extended attachment areas for some of the species which are less given to the aggregating habit. Off both coasts of Long Island, and especially at its northern end locations, can be found bottom sites where many Ascidian species are growing side by side. Inside Cape Hatteras off Beaufort, North Carolina, and off the Georgia coast, as already stated, there are several "live bottom habitats" containing hundreds of intermingled common off-shore specimens of many species. Similar aggregations occur off the southern Florida Keys, and also on the northern coastal shallow sand flats in the Gulf of Mexico south of Panacea and off southern Texas. Figure 5 shows where some of the live bottom habitats have been studied.

As already described, there are several well-defined locations over the edge of the continental shelf and on the continental slope where a few specialized deep water species hold on in the dark and make a meager living from the tiny food particles which a favorable current sheds.

In this chapter the genera and species are listed under the family headings, which are repeated here for convenience. A brief description of each species is given, along with a statement concerning its distribution. These, with a series of line sketches, should aid in identification of specimens to species after they have already been placed into the family groups. With each of the more important species there are included brief notes on the time of maturity and suggestions as to when to collect. The special characters concerning time of budding for colonial species, and any special characteristics concerning development are mentioned. It is usually stated when tadpole larvae are formed, as well as notes on the reduction or loss of the tadpole.

Of course, it often happens that only one or two specimens are found at any one time. These can be identified best by following the method already outlined. Observe the specimen for the ordinal characters and then pass on to the distinctive marks of the family. Here it will be most profitable to run through the species lists in Table III, and then check the descriptions of species which match the specimen. In every specimen being studied the species descriptions should be checked by dissection of one or more individuals.

This chapter gives the complete classification review of all the species of *Ascidiacea* I have found on the Atlantic continental shelf, and the list of species is repeated in the Appendix. This review includes eighty-eight species which are noted under their proper family and subfamily headings. Although each of the family groups has already been illustrated by representative species on large plates from our own fresh specimens, the species

Table III. Complete List of Species of *Ascidiacea* Collected 1968–1974 from the Atlantic Continental Shelf of the United States

The superscript numbers following species' names indicate the average depth in meters at which specimens were found as follows:

			Number found
blank	=	10 meters or less	51
2	=	10 to 100 meters	35
3	=	100 meters or more	2

A total of 88 species were identified by the author from specimens collected or seen in collections of others.

Classification		Species	Figures or Plates
Subphylum	*Tunicata*		
Class	*Ascidiacea*		
Order	*Enterogona*		
Suborder	*Aplousobranchia*		
Family	*Cionidae*	*Ciona intestinalis*	Plate I & Fig. 24
	(Diazonidae)		
	Clavelinidae		
Subfamily	*Clavelininae*	*Clavelina oblonga*	Plate II & Fig. 25
		C. picta	
		C. gigantea^{2}	

Classification	Species	Figures or Plates
Polycitorinae	*Eudistoma olivaceum*	Figs. 11 & 26a
	E. capsulatum[2]	Figs. 26b & c
	E. hepaticum[2]	Fig. 26d
	E. tarponense	
	E. carolinense[2]	Fig. 26e
	Cystodytes dellechiajei[2]	Fig. 26f
Holozoinae	*Distaplia clavata*	Fig. $27a_{1-2}$ & d
	D. bermudensis	Plates III & IV, & Fig. 27b
	D. stylifera	Fig. 27c
Polyclinidae		
(Euherdmaniinae)		
Polyclininae	*Aplidium pallidum*[2]	Figs. 28a & b
	A. glabrum[2]	Fig. 28c
	A. constellatum	Plates V & VI
	A. stellatum	Fig. 28f
	A. pellucidum[2]	Fig. 28d
	A. exile	Fig. 28e
	A. bermudae	
Didemnidae	*Didemnum albidum*[2]	Plate VII & Fig. 29a
	D. candidum[2]	Plate VII & Fig. 29b
	D. vanderhorsti	Fig. 29c
	D. amethysteum	
	Trididemnum tenerum[2]	Fig. 29d
	T. savigni	Plate VIII & Fig. 29e
	T. orbiculatum	Fig. 29f
	Diplosoma macdonaldi	Plate VIII & Fig. 29g
	Lissoclinum aureum[2]	Fig. 29i
	L. fragile	
	Echinoclinum verrilli[2]	Fig. 29h
Phlebobranchia		
Perophoridae	*Perophora viridis*	Figs. 12, 30a & c
	P. bermudensis	Figs. 12 & 30b
	Ecteinascidia turbinata	Figs. 13 & 30e
	E. conklini	
	E. tortugensis	Figs. 13 & 30d
Corellidae		
Rhodosomatinae	*Rhodosoma wigleii* sp. nov.	Figs. 14 & 31a
Corellinae	*Corella borealis*[2]	Figs. 14 & 31b
	Chelysoma macleayanum	Figs. 14, 31c & d
(Hypobythiidae)		
Ascidiidae	*Ascidia prunum*[2]	Figs. 15, 32a & h
	A. callosa[2]	Figs. 32b & i
	A. obliqua[2]	Figs. 32c & j
	A. nigra	Figs. 32e & k
	A. interrupta	Fig. 32d
	A. curvata	Fig. 32f
	A. sp. uncertain (like *A. corelloides*)	Fig. 32g
(Agnesiidae)		
(Octacnemidae)		

(Table III continued)

Classification	Species	Figures or Plates
Pleurogona		
Stolidobranchia		
Pyuridae		
Bolteniinae	*Boltenia ovifera*[2]	Frontispiece & Figs. 33a & b
	B. echinata[2]	Plate IX & Figs. 33c & d
	Halocynthia pyriformis[2]	Figs. 33e, f & g
Pyurinae	*Pyura vittata*	Plate IX & Figs. 34a, b & d_{1-2}
	Microcosmus exasperatus[2]	Fig. $34c_{1-2}$
Heterostigminae	*Cratostigma singulare*	Plate X & Fig. $34e_{1-2}$
Molgulidae		
Molgulinae	*Molgula citrina*	Plate XI & Fig. 36d
	M. complanata	Plate XI & Figs. 36e, 37a & b
	M. manhattensis	Plate XII & Fig. $36a_{1-2}$
	M. robusta	Fig. 36c
	M. provisionalis	Fig. 36b
	M. occidentalis	Fig. 38
	M. siphonalis[2]	Fig. 39a
	M. retortiformis[2]	Plate XII & Fig. $39c_{1-2}$
	M. griffithsii[2]	Fig. 39b
	M. arenata	Plate XII & Figs. 40 & 41
	M. lutulenta[2]	
Eugyrinae	*Bostrichobranchus pilularis*[2]	Plate XIII
	Eugyra arenosa padrensis	
Styelidae		
Botryllinae	*Botryllus schlosseri*	Plate XIV & Fig. $43a_{1-2}$
	B. planus	Fig. 43b
	Botrylloides nigrum	Fig. 43d
	B. aureum	Fig. $43c_{1-2}$
Polyzoinae	*Symplegma viride*[2]	Plate XV & Fig. $43e_{1-2}$
	Polyandrocarpa maxima	Plate XV & Fig. $43g_{1-2}$
	P. floridana	Plate XV
	P. tincta	
Styelinae	*Pelonaia corrugata*[2]	Fig. 44e
	Dicarpa symplex[3]	Figs. 21 & 43f
	Polycarpa fibrosa[2]	Plate XVI & Figs. 18, $42a_{1-2}$ & e
	P. obtecta[2]	Fig. 42b
	P. circumarata	Fig. 42c
	P. albatrossi[3]	
	Cnemidocarpa mollis	Figs. 19 & $42d_{1-2}$
	C. mortenseni[2]	Figs. 19 & 42f
	Styela partita	Figs. 16 & $44a_{1-2}$
	S. plicata	Figs. 17 & $44b_{1-3}$
	S. coriacea[2]	Fig. 44d
	S. atlantica[2]	
	S. clava[2]	
	Dendrodoa pulchella[2]	Fig. 44f
	D. carnea	Figs. 20 & 44g

Note: Four families and one subfamily noted in parentheses are from outside the American shelf area and are included to describe additional evolutionary directions taken by Ascidians.

review includes line sketches of most species seen and identified, arranged in groups. These line drawings are usually made by redrawing published sketches four or five at a time in similar species comparisons. The use of certain redrawn line drawings from Van Name (1945) or Berrill (1950) (with permission) gives greater uniformity than would the use of our own species sketches made from all angles. Van Name was very careful to have line drawings made which showed zooid organization of the particular Ascidian species at about the same angle and at an approximately uniform size, regardless of the magnification required. Our own specimens were used for preparation of each species sketch with the exception of five species for which it was necessary to examine rarer museum specimens.

ASCIDIACEA

Enterogona—Aplousobranchia

Most of the members of this suborder are colonial species of Ascidians in which the elongate zooids are divided into thorax, abdomen, and, in one family, also postabdomen. The gut loop is at the posterior end, or in the postabdomen, and there is a V-shaped heart at the extreme end beyond the gonads. In the *Cionidae* the epicardia in development form two sacs which envelop the viscera above on each side. In a few other families the epicardia form a stolonic tube between the gut and heart. All these species except *Ciona* are viviparous, but in one family the eggs are released into the fleshy tunic and develop there.

Some years ago Berrill (1950) picked out *Ciona* as probably the most primitive living Ascidian species in waters of the northern hemisphere, since, as noted above, in its development the atrial cavity forms from two separate invaginations on the dorsal side. Following this suggestion, Kott (1969) has placed the family to which it belongs with the *Aplousobranchia*. Therefore, as a solitary species *Ciona* becomes also the most primitive of this suborder. I believe that this change in the classification of the *Cionidae* can be justified structurally, and it is followed here. So this group of solitary transparent zooids becomes the first family of *Aplousobranchia*.

Cionidae. Ciona intestinalis is the only widely distributed species (Plate I & Fig. 24).

It forms an elongate tapering zooid attached by the base and with two elongate siphons. There are eight light-sensitive orange spots around its oral siphon and six around the atrial. The mantle shows six longitudinal muscle bands.

The branchial sac has no folds, but shows many internal longitudinal vessels. There are large, comma-shaped papillae at the intersections of the alternate large and small transverse tubes. The dorsal tubercle opens forward, and the dorsal lamina is cut into languets.

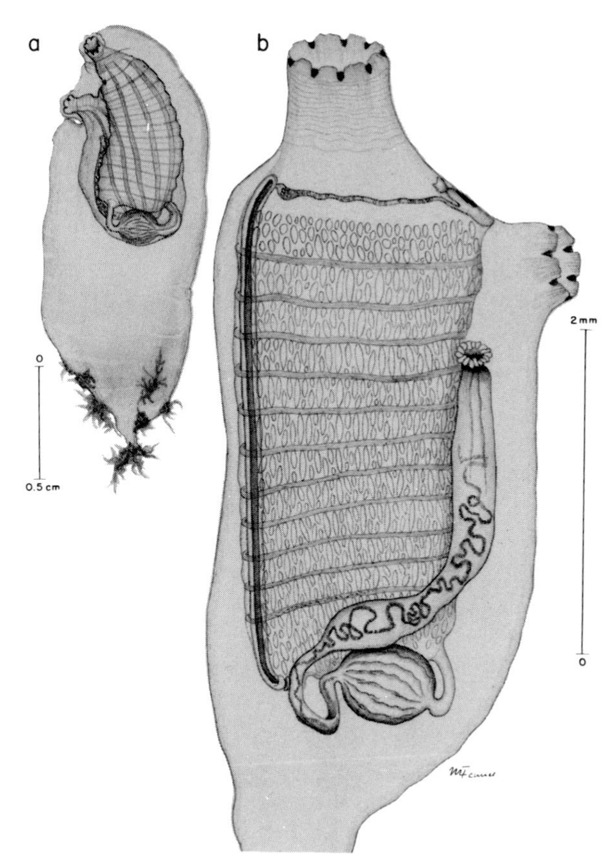

Fig. 24. *Ciona intestinalis.* (a) Small specimen with test. (b) Young, but larger specimen with test removed.

Most of the loop of the digestive tube, esophagus, stomach, and intestine lies ventral to the branchial sac, and the rectum extends left dorsally into the atrial cavity. The gonads are placed inside the intestinal loop, so are on the left side. The large oviduct and smaller sperm duct accompany the rectum. The animals are oviparous, and the eggs are released from mature zooids before dawn. They develop into small, swimming tadpoles after shedding in the morning.

Ciona intestinalis is widely distributed in the northern hemisphere off Europe, from the Mediterranean to the arctic. Off North America in the Atlantic it occurs from Greenland to Cape Cod. It is common in circulating salt water aquaria in Woods Hole and other marine laboratories. In Figure 5 is shown the presence of *Ciona* in the Gulf of Maine and south to New Jersey from dredging records.

Ciona intestinalis has been reported from Suez,

Straits of Magellan, and Australian ports. There is no evidence that such ship-carried specimens have become permanently established. Verrill (1871) and Whiteaves (1900) described a smaller, light-colored variety from off Grand Manan Island called *C. tenella*. Similar varieties may still be found in isolated areas farther south, but there is no constancy in their range. In general, northern sites for *Ciona* on the Atlantic continental shelf appear to be less common than fifty years ago, but it is found at Woods Hole and in harbors south at least to Beaufort, North Carolina.

Clavelinidae is a large family in which buds form from the stolonic vessel. It is separated into three subfamilies. The majority of the species are found in warmer seas south of Cape Hatteras.

Clavelininae. These are species with long clear zooids consisting of an elongated branchial sac and an extended abdomen below. There are at least fifteen rows of stigmata in the branchial sac. When eggs are mature they form a large mass in the atrial cavity and they develop there to tadpoles. During the winter the buds form quiescent statoblasts which grow out as functional zooids in the spring. *Clavelina oblonga* was dredged by the Research Vessel *Asterias* and later by the Research Vessel *Gosnold* from the continental shelf outside of Cape Hatteras, North Carolina. It is common off Sapelo Island, Georgia, and south from there to Florida, the West Indies, and Bermuda (Plate II & Fig. 25). Hartmeyer (1924) reported a possible specimen in the winter statoblast stage from Davis Strait off Greenland. This is one reason for the belief that occasional specimens of this warm water species may wander north of their normal distribution area in exceptional seasons. In 1974 University of New Hampshire collectors found several young specimens of *Clavelina oblonga* in shallow water north of Boston Bay. There are no other records of permanent residence north of Cape Hatteras.

A structurally similar species, *Clavelina picta* has a wide distribution off both coasts of Florida, the West Indies, and Bermuda. It is more brightly colored, pink or carmine, and the zooids form large test masses at their bases. Like the previous species, these *Clavelinas* are intermixed with many corals or gorgonians and Distaplids in shallow water of five to ten meters.

An additional species, *Clavelina gigantea*, has been found to be common off the Florida west coast from Tampa Bay to Apalachee Bay in deeper water. The zooids are regularly encased in the thicker rather tough test which contains a black pigment diffused through the mass. The branchial sac is somewhat more massive than in the commoner *Clavelina* (Plate II).

Polycitorinae. As now described, this small subfamily contains a number of species which are frequently col-

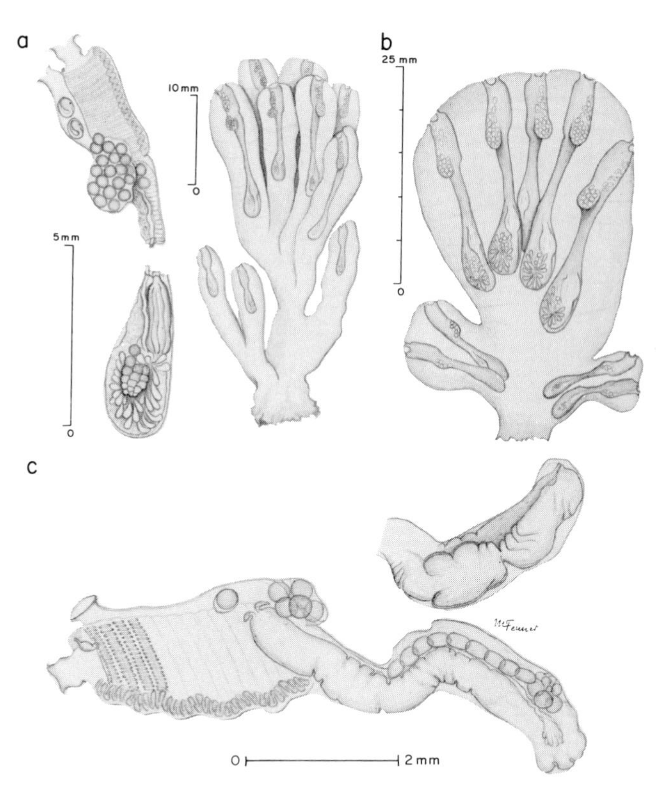

Fig. 25. *Clavelina*. (a) *C. picta*, individual and colony; (b) *C. gigantea* colony; (c) *C. oblonga*.

lected in southern waters. They are massive and rather soft colonies, bright purple, red to green, without spicules, covered with tiny elevations at the openings of the zooids. The zooids are rather long and clearly separated into thorax and abdomen. The thorax contains a wide branchial sac with only three rows of stigmata. The abdomen is long with the reproductive organs at the lower end under the stomach. These belong to the genus *Eudistoma*. Zooids are easily removed from the colonies. They can be stained and mounted and the three rows of stigmata are easily visible.

The *Eudistoma* species may be dredged from water above the continental shelf off Georgia, but are more common off the west coast of Florida and in the Gulf of Mexico south off Padre Island, Texas.

The commonest species is probably *Eudistoma olivaceum* collected in large, rounded, green-grey masses attached to algae at depths of 5 feet in Apalachee Bay, Florida. The zooids can be removed and mounted. They are light colored and about 4 mm long. The masses are about 6 inches across (Figs. 11 & 26a).

Eudistoma capsulatum is quite common in deeper water of 25 meters off the west coast of Florida and south of Texas. R. Defenbaugh of Texas A. and M. University collected the masses from which the draw-

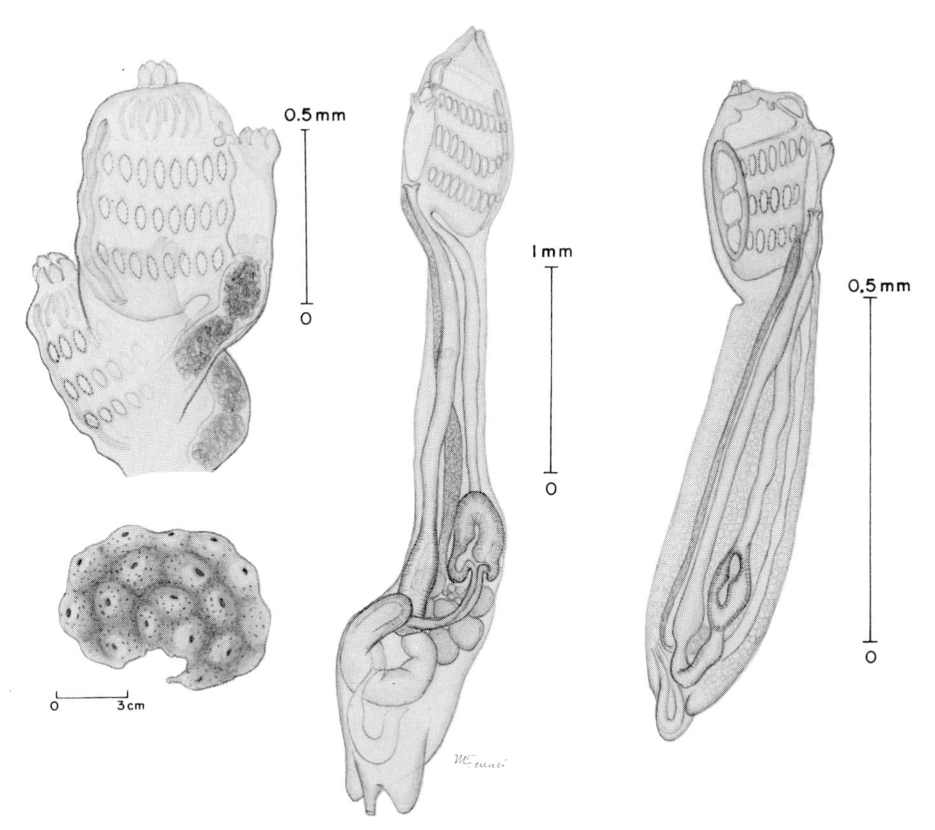

Fig. 26a. *Eudistoma olivaceum.* Branchial bud; colony; two zooids.

ings of the zooids were made. The specimens still retain their violet-gray coloring after eight years. The zooids average 6 mm in length (Figs. 26b & c).

Eudistoma hepaticum is another species widely distributed west of Florida and south of Texas at 20 meter depth. It is colored a deep reddish purple, justifying the species' name, and the color is permanently retained. The zooids of this species too are easily removed and when mounted clearly demonstrate the characteristic three rows of stigmata. Well shown also is the convoluted intestine with the intestinal gland along its outer surface. The gonads are placed at the bottom (Fig. 26d).

Eudistoma tarponense is a shallow water, yellow, encrusting species from the Florida west coast, with smaller zooids.

Eudistoma carolinense forms clumps with many distinct heads all interconnected at the base. The clumps may be 25 mm or larger and are easily recognizable because they are densely crowded with coarse sand grains. Zooids are easily removed. Since they are transparent, the zooids—2 to 3 mm—can be stained and mounted, and show plainly the three rows of stigmata (Fig. 26e).

When collecting *Eudistoma* on the coast of Florida one finds encrusting masses, black or purplish, which feel rough and are very tough. When these masses are sliced open with a scalpel, it can be seen that over each zooid is a tough mass of overlapping disc-shaped spicules which contains four rows of stigmata. This identifies the specimens as belonging to another genus—*Cystodytes dellechiajei* (Fig. 26f). This species is found in shallow water in Apalachee Bay. It is described as occurring in somewhat deeper water off the Pacific coast. This is probably an ancient species, as suggested by the preliminary identification by Francoise Monniot of a Pliocene fossil from Breton as *Cystodytes incrassatus.*

Holozoinae (Plates III & IV) is a distinct subfamily which includes a number of widely distributed species of colonial Ascidians with the rows of stigmata reduced to four. The eggs and tadpole larvae develop in an incubatory sac extending from the mid-dorsal side of the zooid. In the pouch the eggs move outward to the tip, so that the youngest are at the end. Vegetative reproduction comes about by development of buds from the tip of the stolonic epicardium at the lower end of this zooid. The colonies are variable in shape, either capitate or thick encrusting masses. They vary in color in life from yellow to greenish red, and even light blue. On the Atlantic continental shelf there is one rather rare north-

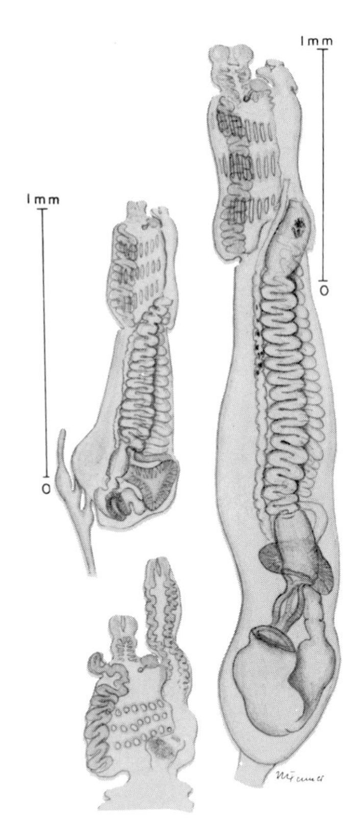

Fig. 26b. *Eudistoma capsulatum.*

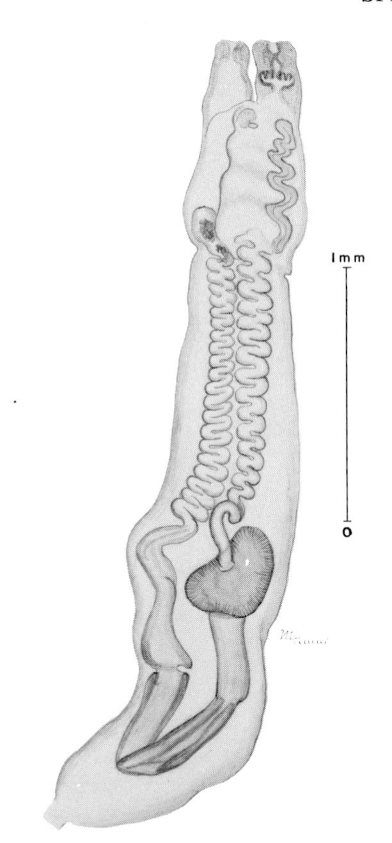

Fig. 26c. *Eudistoma capsulatum.*

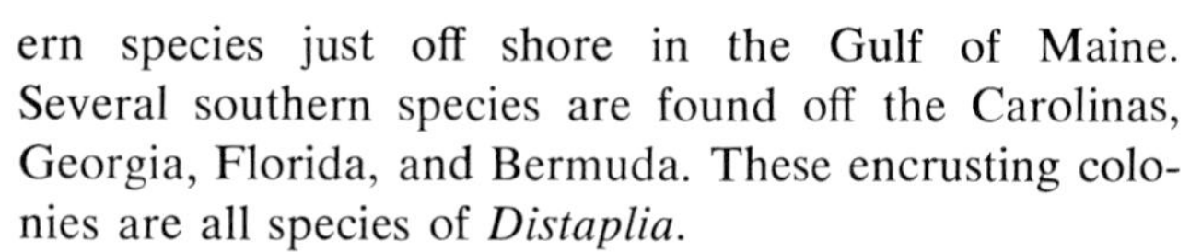

ern species just off shore in the Gulf of Maine. Several southern species are found off the Carolinas, Georgia, Florida, and Bermuda. These encrusting colonies are all species of *Distaplia.*

The northern species is *Distaplia clavata*, recorded from off Eastport and recently (1974) by New Hampshire collectors north of Boston Bay. The colony consists of a dozen stalks about 6 mm high. They are yellowish below, grading into gray-blue above. There are four rows of stigmata, each crossed by a parastigmatic vessel (cf. Figs. $27a_{1-2}$ & d). The stomach is rounded and has fine longitudinal lines on the surface. There is a tubular gland surrounding its upper end, the duct from which joins the stomach.

The gonads of *Distaplia* lie on the right side of the intestinal loop from which oviduct and sperm duct pass to the atrial cavity. The brood pouch, when it is present in mature zooids, passes downward outside the atrial cavity, breaking out from the dorsal side just below the base of the branchial sac. *Distaplia clavata* zooids are often widely separated, with test only at base.

The commonest species on the Atlantic continental shelf is *Distaplia bermudensis* (Plates III & IV, Fig. 27b). It occurs in colonial masses found off Sapelo Island, Georgia, and also off both coasts of Florida. Large colonies appeared at 20-meter depth west of Tampa Bay in the *Hourglass* samples. The structure of the zooids is very similar to *Distaplia clavata*, but sturdier, and the colonies are larger, with heavy masses running to 20 mm. In Apalachee Bay, colonies of many bright colors are found on algae or on other Ascidians at depths of 10 to 20 meters. Such masses are usually of bright red or pink, blue, brown, or green. Such striking colors are unique among colonial Ascidians. This species was described from Bermuda and it occurs throughout southern and West Indian waters.

The species of *Distaplia* found along the Florida west coast have not been carefully studied. In addition to the bright colored varieties, at least two additional distinct species have been distinguished. One is *Distaplia stylifera* (Fig. 27c), which is an Indian Ocean form which ranges to Australia as well as into the West Indies. These form small mushroom-like beads colored yellowish brown. Zooids differ from *D. bermudensis* in the growth of an extended sac from the abdomen containing the reproductive organs. In addition, a brood pouch is extruded from the abdomen in which the eggs develop.

Still another species may have appeared in material dredged from 10 meters depth in Apalachee Bay. This is another of the "live bottom habitats" in which many different species are crowded together. It is a *Distaplia* species or subspecies not yet described, in which the

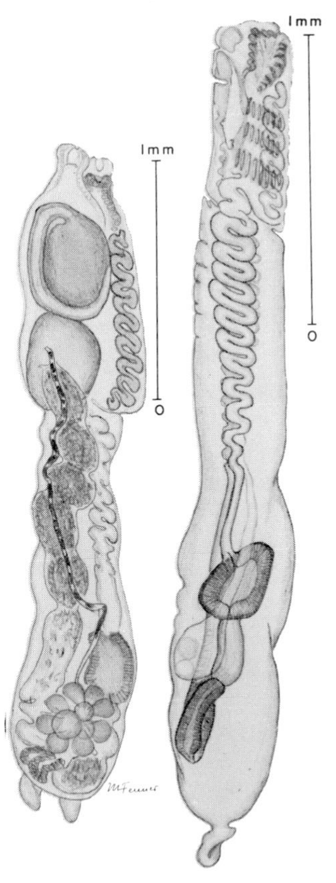

Fig. 26d. *Eudistoma hepaticum.*

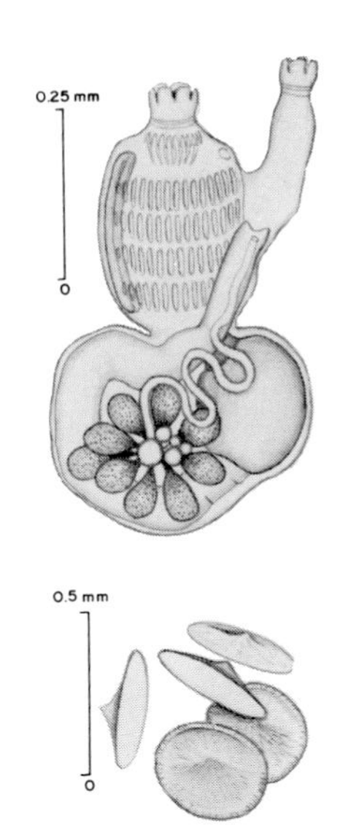

Fig. 26f. *Cystodytes dellechiajei* (after Van Name).

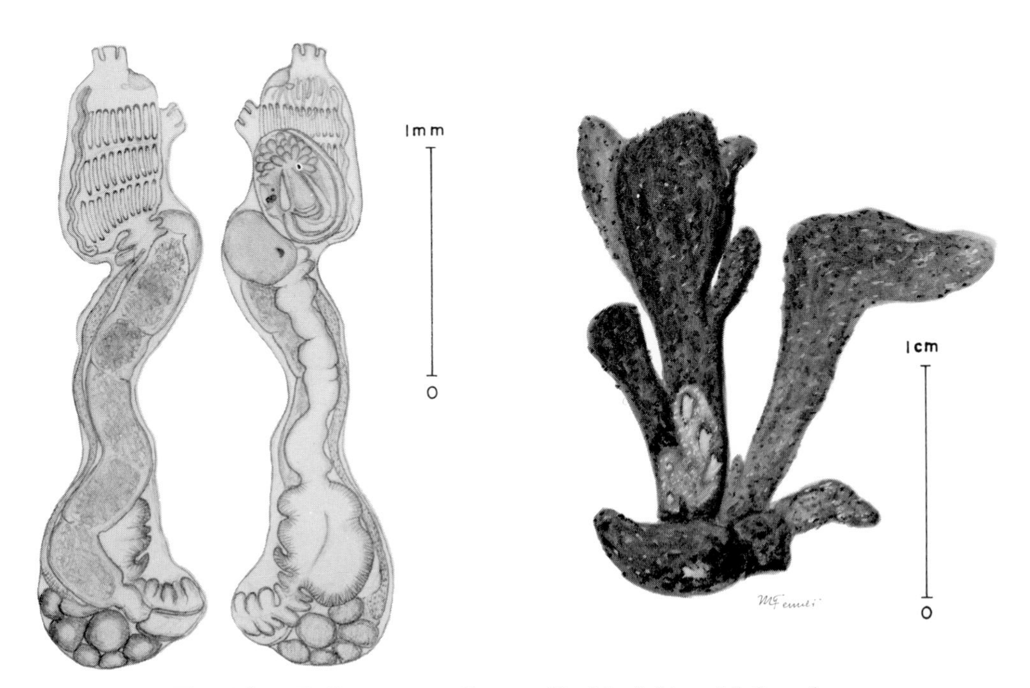

Fig. 26e. *Eudistoma carolinense.* Zooid, right and left; colony.

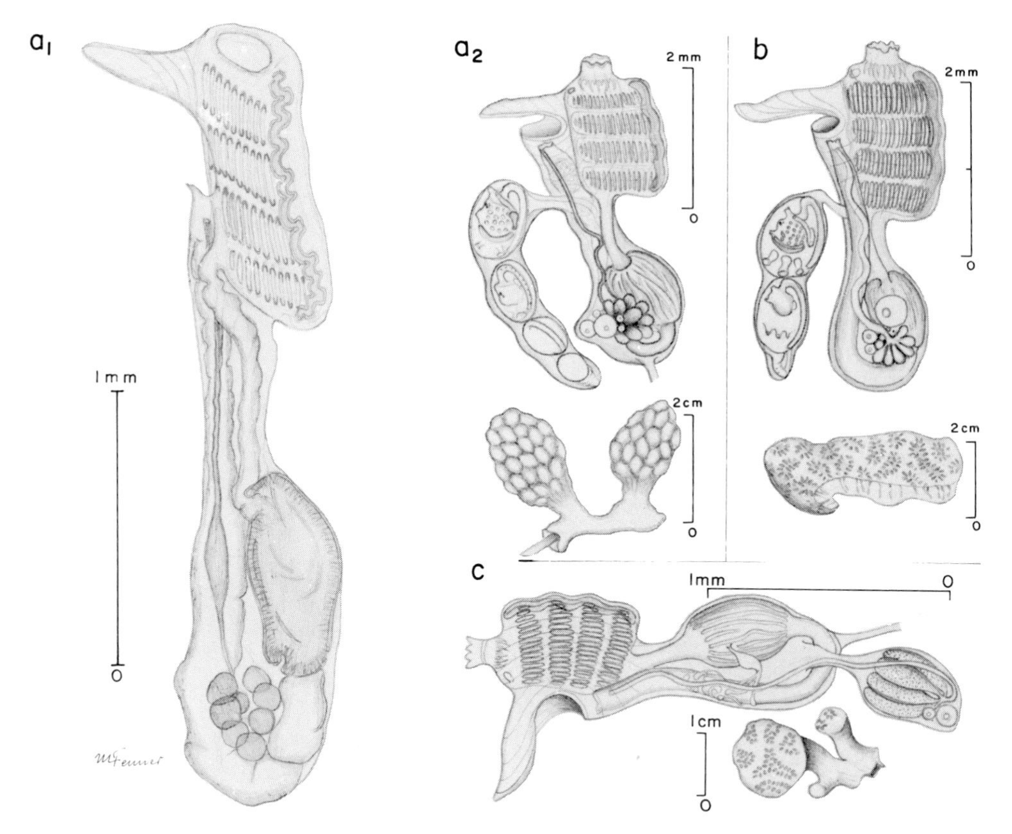

Fig. 27a–c. *Distaplia clavata*: (a_1) zooid, (a_2) zooid with brood pouch, colony. (b) *D. bermudensis*: zooid with brood pouch, colony. (c) *D. stylifera*: zooid, colony (a_2, b, and c after Van Name).

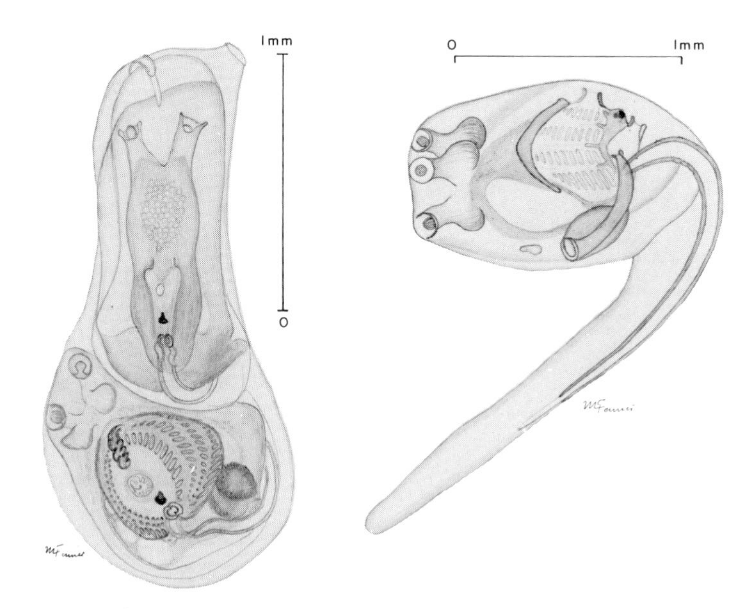

Fig. 27d. *Distaplia clavata*: brood pouch, tadpole.

esophagus originates from the inner corner of the branchial sac. This species forms colonies ordinarily colored pink or brick red. These specimens need further study and are as yet undescribed, presently called *Distaplia* sp.

Polyclinidae. This large family is made up of colonies containing zooids imbedded in the matrix and forming systems. The individual zooids are divided into thorax, abdomen, and postabdomen. Vegetative reproduction occurs by constriction or annulation of the postabdomen. Eggs develop into complete tadpoles released from the atrial siphon.

Polyclininae. This is a subfamily of colonial species found all over the world. They are especially numerous in the Gulf of Maine and around Cape Cod. The colonies are often massive and in life their bright color shows through the surface dull or gray tone. The semitransparent zooids are 4 to 6 mm thick, but the masses enlarge by budding. Even in colonies of the same species there may be red, orange, purple, or light gray masses. The zooids are divided into thorax, abdomen, and often postabdomen. Only in this family is the heart a long tube bent in a V behind the gonads at the extreme end of the postabdomen. The gut loop is posterior to the branchial sac in the abdomen. The gonads are arranged along the sperm duct in the often very long postabdomen.

These Polyclinids are viviparous, with eggs released into the peribranchial or atrial cavity which develop there into complete tadpole larvae with visual sense organ and statolith. Each segmental mass forms a new zooid which migrates to the surface of the common test or tunic.

There is one major genus which is distributed at shallow or moderate depths from arctic to tropical marine areas. This is *Aplidium*, named by Savigny in 1816 to include species broadly attached and with ten or more rows of stigmata in the branchial sac, and a long postabdomen. A larger number of species were designated *Amaroucium* by Milne Edwards over a hundred years ago because they were elevated in capitate form and testes were serially repeated. At most, the latter is a subgenus of *Aplidium* and should be given in parentheses following the genus name. In general, it seems better to substitute the older genus name with each species and omit the subgenus. That practice of many specialists is being followed here, although both terms will be used when the latter species are first mentioned.

Two northern species are found widely scattered in the Gulf of Maine. They are: *Aplidium* (*Amaroucium*) *pallidum* and *Aplidium* (*Amaroucium*) *glabrum.*

Aplidium pallidum was described by A. E. Verrill in 1873. It is a northern European species which has been found in Davis Strait, east of Greenland, down to the

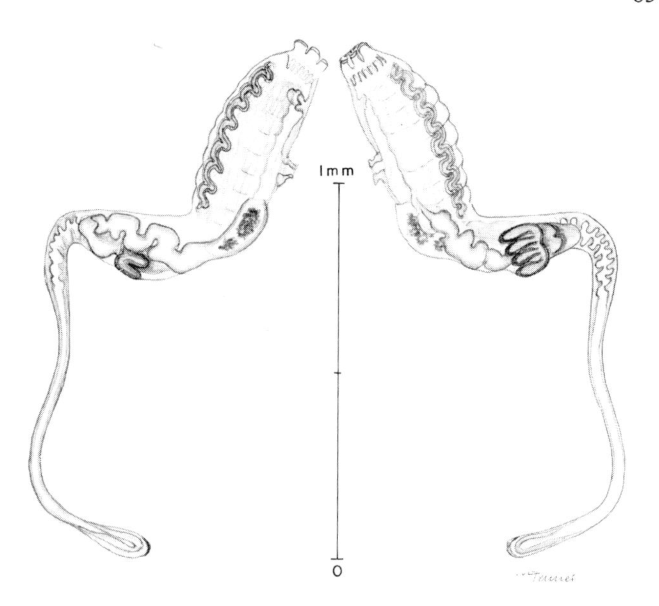

Fig. 28a. *Aplidium pallidum*: immature zooid from bud.

Gulf of St. Lawrence. It occurs in deeper water in the Gulf of Maine and south of Cape Cod to off Block Island. It prefers sandy bottoms where there are some small stones for attachment at depths up to about 150 meters. *A. pallidum* forms low, round-topped masses, flattened above and about 25 mm across. These are grayish and usually covered with fine sand. Each head is made up of several irregular systems of eight to ten individuals around one cloacal opening. The short irregular zooids can often be seen through the thin test. They have a short postabdomen.

When zooids are teased out of the mass and examined with the microscope it can be seen that the short zooids of about 3.5 mm terminate in a six-lobed oral siphon with a lateral atrial opening. There are seven to ten rows of stigmata in the branchial sac. The abdomen and postabdomen are short. The ovaries lie directly behind the intestinal loop and the testes are annular masses below (Plate VIII & Figs. 28a & b).

Another widespread northern species which forms small round-topped capitate masses is the circumpolar *Aplidium glabrum.* These are common in the Bay of Fundy, off Eastport, and on Georges Bank. Often these 20-mm heads contain only one system of zooids with a single central opening. Occasionally the masses are larger, up to 50 mm or more. The colonies appear light blue-gray and they are usually full of sand above.

The yellowish zooids when teased out are longer than those of *A. pallidum*, averaging 5 mm, with long thin postabdomens. There is a large atrial languet. The stomach has ten to twelve ridges longitudinally. Sometimes each of these last two species may appear with a small number of low encrusting heads, but examination

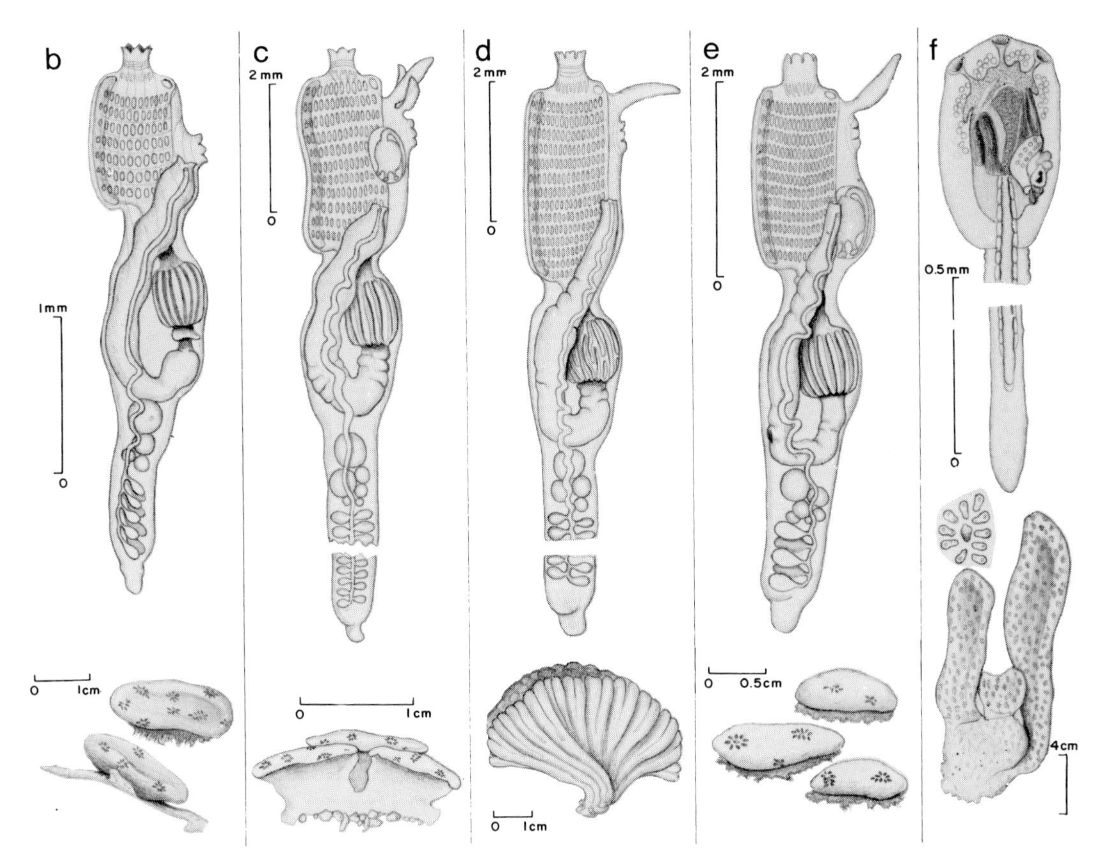

Fig. 28b–f. *Aplidium* species: single zooids and colonies. (b) *A. pallidum*; (c) *A. glabrum*; (d) *A. pellucidum*; (e) *A. exile* (after Van Name); (f) *A. stellatum*, larva and colony (after Van Name).

of the zooids dissected out of the mass will separate them quickly. These zooids have twelve or more rows of stigmata (Fig. 28c).

There are three species of *Aplidium* (*Amaroucium*) which are well known to marine zoologists in the region south of Cape Cod in Vineyard Sound and Woods Hole. In addition to furnishing specimens for study of adult morphology, they release swimming tadpoles in the morning during most of the summer. These species are *Aplidium stellatum*, *Aplidium pellucidum*, and *Aplidium constellatum.*

The first of these three, *A. stellatum*, is found sporadically in water of medium depth the whole length of the North American Atlantic continental shelf from Cape Cod to Florida. It is easily identified by its thick salmon- or flesh-colored masses, 3 inches wide and up to 3 feet or more long from the point of attachment. This is the best known "sea pork," which though tough and indigestible is sometimes bitten off by sharks, skates, and fish. An enlarged figure of the tadpole larva is shown above the double-pointed colony (Fig. 28f). The zooids when removed are rather similar to those of *A. glabrum* and have usually twelve rows of stigmata. The stomach has ten to twelve well-defined ridges.

The second species, *A. pellucidum*, was first described from material found off Point Judith, Rhode Island. This species appears to be made of heavy sand-covered columns which at first sight look like sponges (Fig. 28d). It was the first of these *Aplidium* zooids to be described, and it is difficult to understand why Leidy gave these colonies such an inappropriate name.

Each column surrounds a group of narrow zooids with a central cloaca. There are fifteen or more rows of stigmata in this species. The orange, rounded stomach is covered with about twenty longitudinal ridges. This species is common on the deeper bottoms of Vineyard Sound off Woods Hole, but it is occasionally picked up from Long Island Sound south to Georgia, Florida, and off Florida's west coast.

The third species, probably the best known of all, is *Aplidium constellatum*. It is found mixed with sponges, coelenterates, and echinoderms over gravelly masses in about 10 meters of water off Nobska Light in Vineyard Sound, and in many similar live bottom areas throughout its range. It is known from Casco Bay, Gloucester, Buzzards Bay, Long Island Sound, and occasionally in sheltered areas south to off Beaufort, North Carolina; Sapelo Sound, Georgia; and both coasts of Florida (Plates V & VI).

Colonies in Vineyard Sound are often masses 25 mm thick and 75 mm or larger across, attached by a narrower base. In color the masses range from orange or salmon to dull gray. When they are removed from the test and examined the zooids measure 5 mm, and the postabdomen may be 10 to 15 mm or more. There is a large atrial languet and there are twelve rows of stigmata in the branchial sac. The stomach is finely ridged with twenty or more lines. There are a series of dorsal languets to the left of the dorsal lamina. In mid-summer when zooids are mature the atrial cavity may show a succession of developing eggs leading to fully formed tadpole larvae. The latter are released from the living zooids in the morning and swim actively for several hours.

In addition to *Aplidium constellatum*, there are at least two more common southern or West Indian species found off the Georgia and Florida coastal marine areas. One is *Aplidium exile*, which forms yellowish or greenish, thin encrusting masses in shallow water (Fig. 28e). The zooids are slender and upright. There are usually twelve rows of stigmata in the branchial sac and many fine ridges on the stomach. Eggs may develop to the tadpole stage within the branchial sac. Another common species of Bermuda, found originally by Van Name off Georgia and Florida, is *Aplidium bermudae.* It forms tough brownish masses which drift on to the sand at Sapelo Island, Georgia. Such masses may be up to 18 inches across. Specimens may be found encrusting shell to 10 meters depth to a half mile off shore. Zooids are similar to *A. glabrum*, but much more tough and in larger clumps.

One other member of this family has been found infrequently in the northern portion of the Gulf of Maine off Mt. Desert Island. This is *Synoicum pulmonaria.* It is a smooth gray colony elevated on a short peduncle. There are in each mass five or more zooids arranged about a common cloaca. The atrial opening is at the end of a short funnel-shaped tube. Eggs from the postabdomen develop into tadpoles in the atrial cavity. This beautiful species of northern European waters is very seldom collected here. A specimen in the National Museum was found off Sable Island, Nova Scotia. This author has never collected it.

Didemnidae. One important world-wide family of *Aplousobranchia* remains to be described. In shallow or in deep water are found tough, whitish (2–3 mm), encrusting layers. With a hand lens it is seen that the colony contains numerous stellate calcareous spicules which produce a chalky, hardened surface. Atrial siphons open into a cloacal system. The small zooids are divided into thorax and abdomen and are arranged irregularly in the encrusting layer. A muscular retractor spike of variable length extends downward out into the tunic from the posterior end of the thorax in many species (Plate VII).

The branchial sac of *Didemnum* has four rows of stigmata, or in *Trididemnum* only three, with no longitudinal folds. The heart is a straight tube along the lower margin of the intestine. The epicardium consists of two cell masses below the esophagus. The gonads are within the gut loop. The ovary appears as one or two eggs only and the testis is a rounded mass. In *Didemnum* the proximal sperm duct is regularly coiled over the outside of the testis, but in *Diplosoma* it is straight (Fig. 29a–i).

The eggs are released into cavities within the test where they develop into tadpoles. These larvae are as large as the adult zooids. Their tails twitch briefly, but they may develop to mature zooids within the test.

"Pyloric budding" in the *Didemnidae* is an extraordinary process, involving the secondary joining of two separate buds formed at the base of the thorax in a functional zooid. The anterior bud gives rise to the abdominal portion of a new zooid and the posterior bud to the thoracic portion. These two separately forming buds may unite to form a single new zooid, or each bud may itself later form the remainder.

The *Didemnidae* are an important and highly specialized group of Ascidians found in all seas. They need much more careful study with experimental cultivation and specimens mounted at different stages of colony formation. Such study might yield important data on the physiology of the Aplousobranchs and even on the evolution of this whole class of animals. The chief deterrent to extensive study is the small size of the zooids and the difficulty in separating and mounting uninjured specimens. Ralph A. Lewin of the Scripps Institution of Oceanography has described Didemnid species which carry green symbiotic algae prokaryotically.

Didemnum albidum is the common cream-white encrusting species of the northeastern continental shelf (Plate VII & Fig. 29a). The masses are seldom more than 50 mm across and are usually chalky white, but they may be yellow or orange-pink. The colonies feel gritty to the touch, and the spicules appear as lumpy, rounded masses (± 0.05 mm) when a bit of colony is examined under the microscope. The oral siphon extends dorsally and the atrial opening is lateral with a long anterior languet. There are four rows of stigmata and a very thick endostyle. The two-lobed testis is at the posterior end and it is overlaid by six to eight turns of the coiled sperm duct. This is a circumpolar species

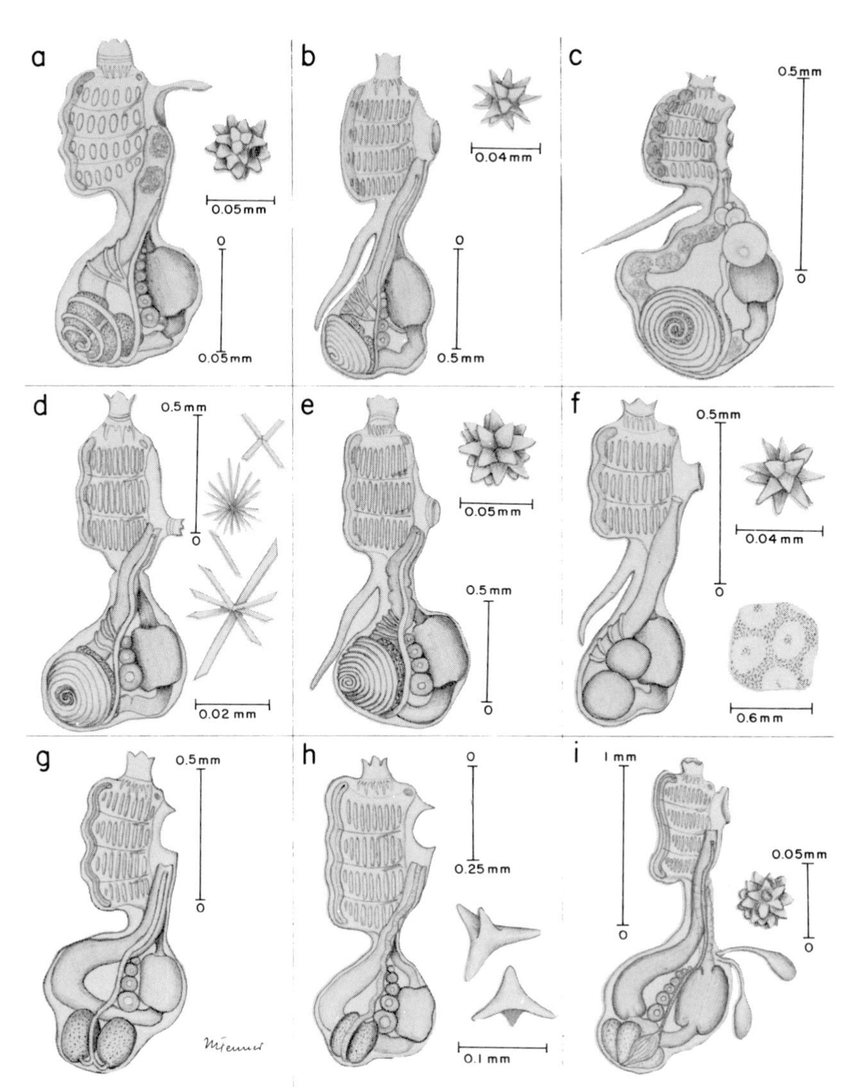

Fig. 29a–i. *Didemnidae* zooids with spicules (after Van Name). (a) *Didemnum albidum*; (b) *D. candidum*; (c) *D. vanderhorsti*; (d) *Trididemnum tenerum*; (e) *T. savignii*; (f) *T. orbiculatum*; (g) *Diplosoma macdonaldi*; (h) *Echinoclinum verrilli*; (i) *Lissoclinum aureum.*

found at 10 to 50 meters from northern Europe, both coasts of Greenland, eastern North America, south to Cape Cod, north of North America to the Bering Sea. It is not found in the Pacific. Its white masses on seaweed or other Ascidians make it one of the best known and the commonest of northerly distributed Ascidian species. There are many local varieties.

Didemnum candidum is a warmer water species which is even more widely distributed than *D. albidum* (Plate VII & Fig. 29b). It is found from a little north of Cape Cod south to Florida, Bermuda, and at least to the coast of Brazil. On the opposite side of the Atlantic, *D. candidum* occurs from Africa to Europe, and also in the Red Sea and the Indian Ocean. Although it is seldom picked up in the Pacific, there are records from southern California. It may be found just outside the tidal area and out to the gravel on the shelf 20 miles from shore. So this closely attached colonial Ascidian is one of the most widely distributed Ascidian species in the world.

D. candidum can easily be distinguished from *D. albidum* on microscopic examination by the absence of an atrial languet, the presence of a spine-like muscular retractor process, and the numerous fine stellate-shaped spicules. There are several local varieties with unusual spicule shapes.

Other less common species off the west coast of Florida are *Didemnum vanderhorstii* (Fig. 29c), with very few spicules, and the red-purple *Didemnum amethysteum*, with the male gonad divided into several separate lobes.

There are many species from southern coasts, but only *Trididemnum tenerum* is found in the Gulf of Maine and farther north (Fig. 29d). Records of Verrill suggest that it was commoner a hundred years ago than it is now. It is usually found in deeper water, 10 meters or more, as small grey masses. In contrast to *Didemnum*, this genus has only three rows of stigmata in the branchial sac. The atrial siphon forms an inverted funnel-like tube toward the lower level of the thorax and a languet is lacking. There are usually many elongate stellate spicules in the test. Like *Didemnum*, the sperm duct is coiled at the proximate end over the testis.

Other *Trididemnum* species are found in southern waters. *Trididemnum savignii* is found from off Sapelo Island, Georgia, all around both coasts of Florida, off Padre Island, Texas, and the West Indies (Plate VIII & Fig. 29e). It forms large distinctive masses in shallow water. The stellate spicules are large and the colonies often show a mottled or spotted coloration because the spicules are bunched or clumped. In addition, dark pigment cells may form spots, or are diffused over the whole colony.

Another species occasionally found off southern coasts on the underside of stones close to shore forms small slate-colored masses. It has a honeycomb appearance because spicules are fewer above zooids and more numerous between the zooids. This is *Trididemnum orbiculatum*, and it differs from all the others in lacking a coiled sperm duct (Fig. 29f).

Diplosoma macdonaldi is a widely distributed shallow water Didemnid found from the Georgia coastal area south around both Florida coasts, up to Apalachee Bay, also off many West Indian Islands. It has four rows of stigmata and a large atrial opening without a languet. It is chiefly distinguished from other members of the family in that it contains no spicules and is a rather transparent encrusting colony. Zooids are usually visible through the test, with black pigment in the mantle around the test, but they are light colored, with orange stomachs. The reproductive organs are different from the opaque Didemnids in that the sperm duct does not coil over the testis (Plate VIII & Fig. 29g).

Lissoclinum aureum is a less common member of the family in northeast American waters (Fig. 29i). Its yellowish gray masses encrusting rocks or larger Ascidians average 25 mm across. There are usually fine, rounded masses of stellate spicules. The branchial sac has four rows of stigmata. The atrial opening is at the dorsal end and bears a short languet. The sperm duct is straight, as in *Diplosoma*, not coiled over the testis. This species can be found in shallow water close to shore at Eastport, Maine, but at greater depths on Georges Bank east of Cape Cod and off Martha's Vineyard. Also found in southern Florida, is a common West Indian species, *Lissoclinum fragile*. This has occasional white spicules. The zooids are gray or pale red.

A related species, *Echinoclinum verrilli* was called a separate genus by Van Name (Fig. 29h). The colonies are flat, encrusting but 5 to 6 mm thick. It is found off the Florida west coast from Tortugas Island north to Apalachee Bay at depths of 5 to 20 meters. It is found also off Bermuda and off many West Indian islands. Zooids are arranged in patterns in the reddish brown masses. Each zooid is enclosed at the base by many very large tetrahedral spicules, often up to 0.1 mm across, forming an efficient protective armor.

Enterogona—Phlebobranchia

This, the second largest suborder of *Ascidiacea*, contains solitary species for the most part, but includes one family with zooids attached by branching stolons in large plant-like colonies. The body is not separated into thorax and abdomen. The gut loop is usually beside the branchial sac on either the right or the left side.

The *Perophoridae* form a small family of rather primitive Ascidians related to the *Ascidiidae*, which are more common in southern waters from Cape Hatteras to Florida and the West Indies. *Perophora viridis* has been collected in Doughboy Sound in shore from Sapelo Island, Georgia. Several other species and the much larger *Ecteinascidia* are found along the west coast of Florida, in the Gulf of Mexico, off the West Indian Islands, and Bermuda. The movement of the family to eastern North American waters apparently occurred in south temperate or tropical zones with later migration northward.

Perophoridae is represented in the northeast by only one species, *Perophora viridis* (Figs. 12, 30a & c). These are small zooids, up to 3 mm. On the outer wall appear surface buds which grow out to form a branch or stolon containing a blood vessel. At intervals, buds along the stolons give rise to a series of zooids in a branching colony with an interconnecting blood circulation. The branchial sac has four or five rows of stigmata and the transverse vessels have some inwardly projecting papillae uniting at their ends to form supplementary longitudinal vessels. The heart is a tube which can be seen pulsating along the posterior part of the branchial sac. In living zooids these pulsations can be seen to stop and reverse every few seconds.

The stomach and intestine form a loop on the left side of the branchial sac and the gonads lie within the loop. The sperm duct runs along the rectum. *Perophora* zooids mature during the late spring and summer, and tadpole larvae released in the morning become attached

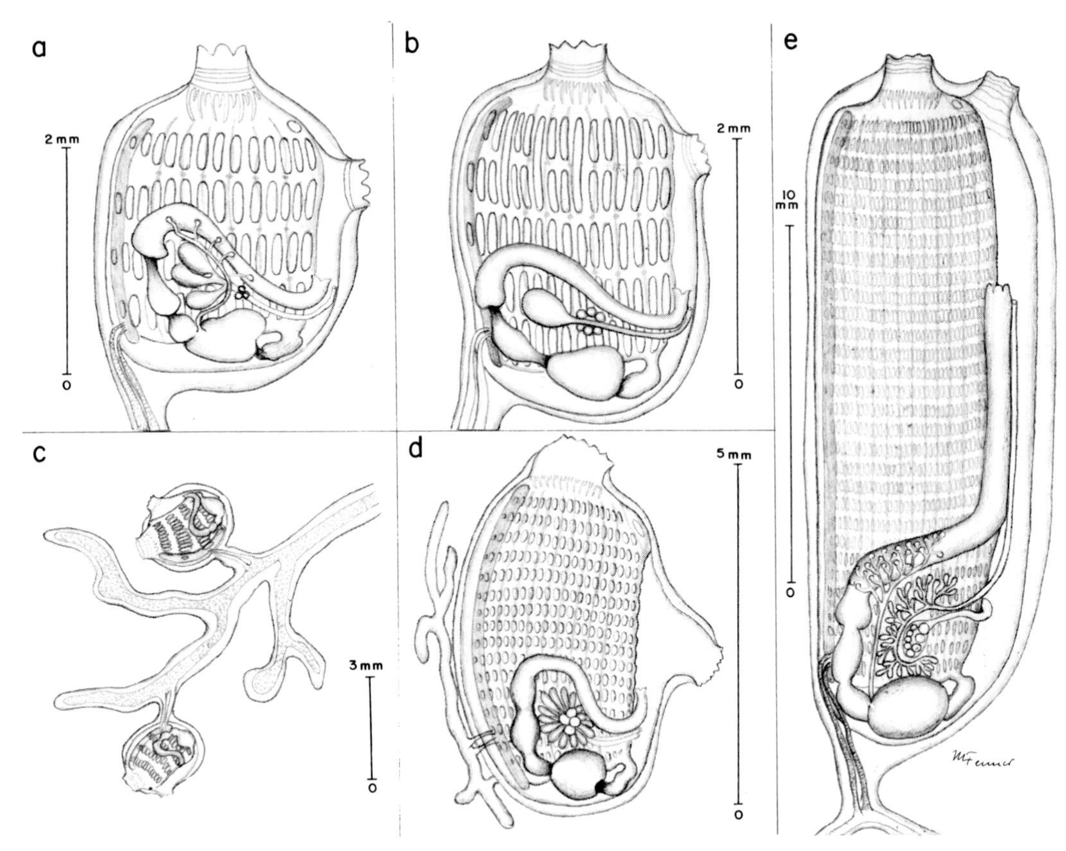

Fig. 30a–e. (a) *Perophora viridis*; (b) *P. bermudensis*; (c) *P. viridis* buds; (d) *Ecteinascidia tortugensis*; (e) *E. turbinata.*

and are functional small zooids by nightfall. *Perophora viridis* of America is similar to *P. listeri* of Europe, and is a world-wide species off temperate shores in shallow water.

Another species, *Perophora bermudensis* (Figs. 12 & 30b), was described from Bermuda, but it is widely distributed along both coasts of Florida and off islands of the West Indies, including Puerto Rico. The zooids are about one-fourth larger and show five rows of stigmata. During the few weeks in summer when they are sexually mature, one zooid may enclose four or five eggs.

The family includes the larger genus *Ecteinascidia*, which is widely distributed attached to sea weeds, mangrove roots, and stones close to shore along Florida coasts. They have been found off the Bahamas and Jamaica, so these are common West Indian species.

Ecteinascidia turbinata forms large clumps in which the elongated zooids are connected by stolons attached to a firm base (Figs. 13 & 30e). Zooids may be 15 to 20 mm long with twenty to thirty rows of stigmata. There are incompletely interconnected longitudinal blood vessels. The esophagus passes to the ventral sac-like stomach, and the intestinal loop bends along the left side to open in the atrial chamber two-thirds of the way toward the atrial siphon. Branching intestinal glands cover the lower intestinal loop and reenter the intestine farther up. The reproductive organs lie in the intestine loop and are mature in mid-summer. Zooids are often bright orange.

At least two additional species of *Ecteinascidia* are common at the Tortugas Islands and perhaps farther south. *Ecteinascidia conklini* is shorter than the previous species. *Ecteinascidia tortugensis* (Figs. 13 & 30d) lies on its side with the atrial siphon two-thirds of the way toward the rear. It is a yellow-green species common at several of the Tortugas Islands.

Corellidae. This is a small Ascidian family which is of interest because its rather unusual structure has some evolutionary implications. Unlike most of the families, its species are found only in the northeastern portion of the North American continental shelf. Also, unlike other Ascidian families, some of the species of *Corellidae* have a gut loop which bends ventrally and then passes dorsally on the right side of the branchial sac (Fig. 14). In one subfamily, *Rhodosomatinae*, the stigmata are straight. In the other, *Corellinae*, they are

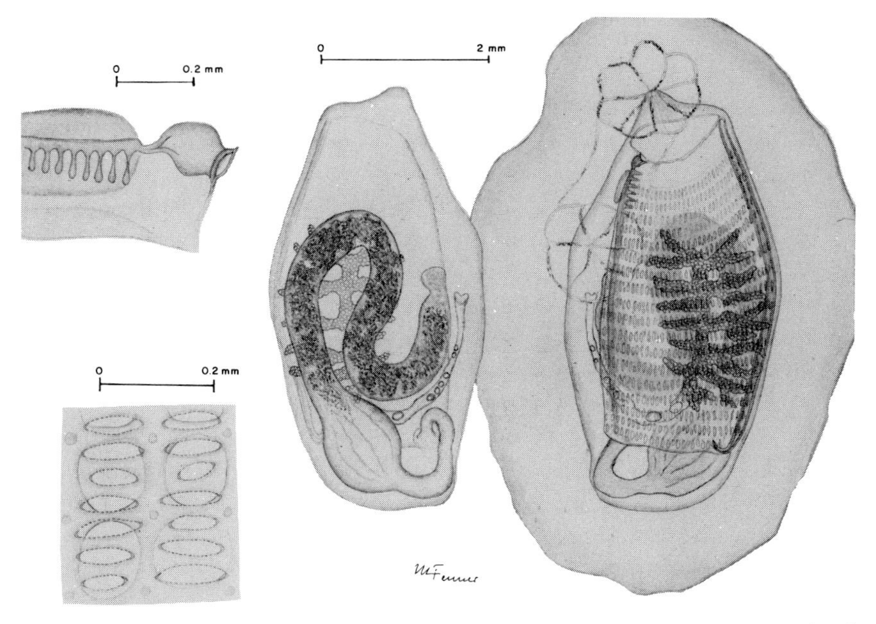

Fig. 31a. *Rhodosoma wigleii* sp. nov. Adults, right and left; details of dorsal tubercle and branchial sac.

partly or completely spiralized. This situation suggests that the enlargement of the stigmata by forming spirals began early in the evolutionary history of the *Ascidiacea*.

The subfamily *Rhodosomatinae* includes a species in which both of the siphon openings are partly covered by a fold of the test or tunic which protects them. A world-wide occurring species, *Rhodosoma turcicum* has been described from waters of the West Indies, but it has not been collected by this investigator. Farther north a number of specimens have been collected of one or more smaller but distinctive species of *Rhodosoma*. They grow on stones at a depth of about 10 meters about one-half mile south of Cape Ann. One of these is shown in Figures 14 and 31a. It has no lid over the siphons, but it has the right-hand loop of the gut and a large intestinal gland. It is noteworthy that the stigmata are straight. It is called *Rhodosoma wigleii*, species novum, after Dr. R. Wigley of the Benthic Laboratory, Northern Marine Fisheries, Woods Hole, Massachusetts. This new *Rhodosoma* species is unlike any other on the Atlantic continental shelf, and it suggests interrelations between *Corellidae* and *Ascidiidae*.

Still another group of specimens also found southeast of Cape Ann, may be still another new variety. It is called tentatively *Rhodosoma*. It resembles *Ascidia corelloides* and has the gut loop on the left side. Both these species will be described separately.

Corellinae. Two species have been collected belonging to this subfamily, one off Cape Ann, and one in the Gulf of Maine south of the Bay of Fundy. These also are collected infrequently in the area. One is *Corella borealis*, a small species found attached in the sand in moderately deep water flowing in shore in the Gulf of Maine from Greenland as far south as Cape Ann (Figs. 14 & 31b). It is an arctic species found in European waters north to Spitzbergen, off both coasts farther south. The zooid is smooth, transparent, often pink-colored. The branchial sac has a dorsal lamina divided into languets. It shows no folds but rather a series of squares formed by regular transverse and longitudinal vessels. Within each is a regular spiral formed by turns of one or two stigmata. The spirals of contiguous squares turn in reverse directions. In addition, there are supplementary internal longitudinal vessels formed by interconnection of papillae extending in from the transverse vessels. The loop of the digestive tube lies unexpectedly on the right side. This, along with the rectum, is overlaid by ovary and testis. Besides migrating southward along the eastern Atlantic continental shelf, this rare and distinctive species of the North American coast is related to others in the Pacific, the West Indies, and even to still others in circumpolar Antarctica.

It is customary to include in this subfamily a related genus of which there is one Atlantic and several Pacific species. The Gulf of Maine species is *Chelysoma macleayanum* (Figs. 14, 31c & d). This is a circumpolar arctic species which ranges southward into the Gulf of Maine as far as Cape Ann. It is a low, flattened, oval zooid, about 15 mm long—larger in the north—firmly attached by the ventral surface to stone or shell. The

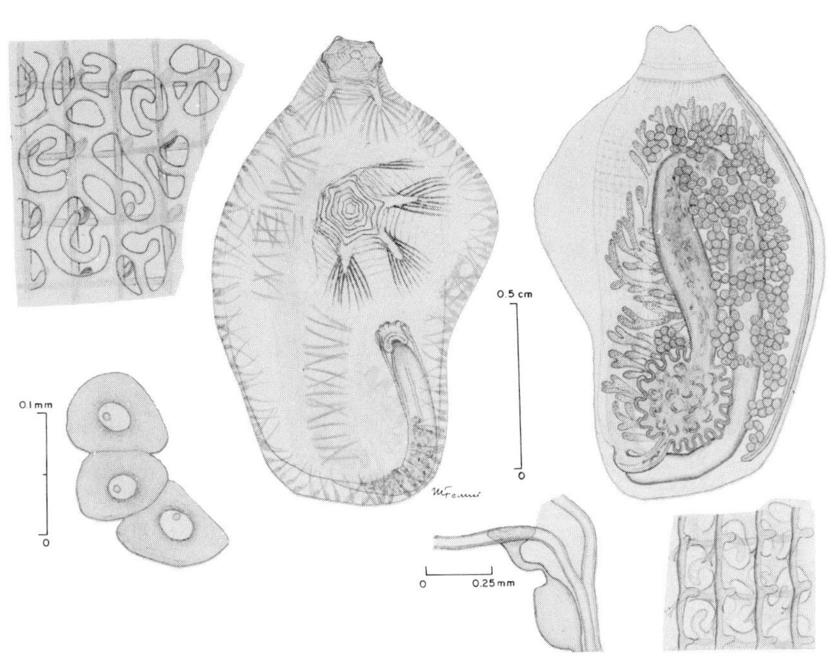

Fig. 31b. *Corella borealis*. Adult, right and left; enlargements of branchial sac, eggs, dorsal tubercle.

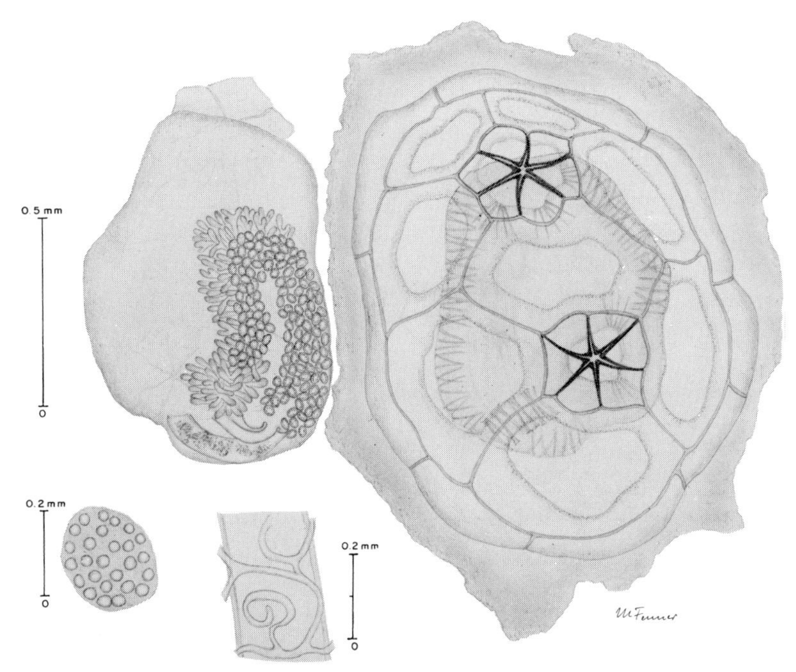

Fig. 31c. *Chelyosoma macleayanum*. Plates of test; gonad; details of egg and branchial sac.

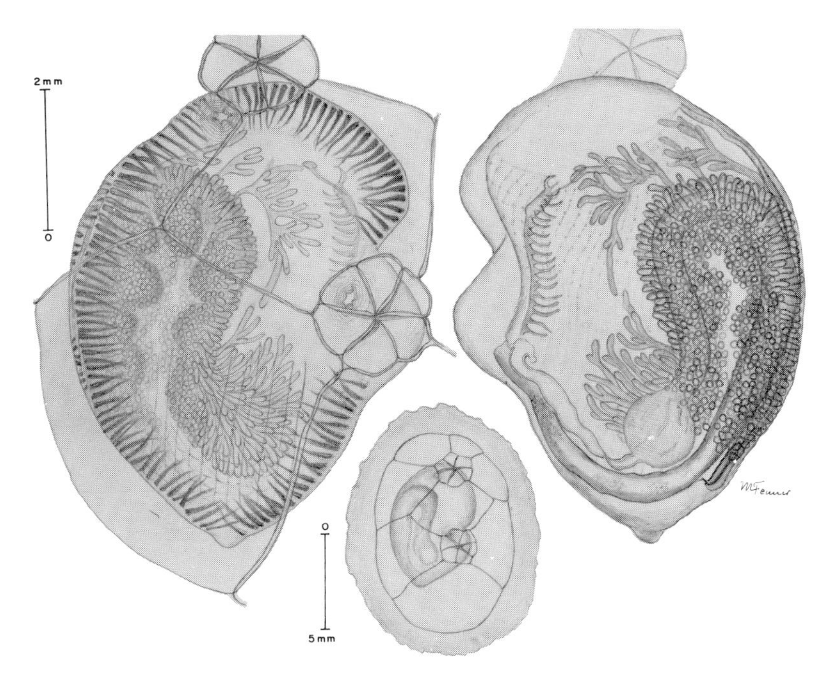

Fig. 31d. *Chelyosoma macleayanum.* Zooid, right and left; orientation within test.

dorsal surface is covered by horny plates, forming a concentric series about each siphon. Each horny plate may show concentric lines of growth which may be used to estimate the age of the specimen (Huntsman 1921). The test is transparent or may be yellowish in older specimens.

When the plates are removed, the branchial sac can be seen with the stigmata forming spirals like *Corella* in a chessboard arrangement. The intestinal loop lies on the right and ventrally. The gonads are spread over the intestine.

These four Corellid Ascidians must come from a very ancient line. Their present distribution suggests migration not only from the arctic but perhaps also from the south, even from the antarctic. One fossil *Chelyosoma*-like specimen from Permian of Sicily, shows that this family was widespread in the warm period of the late Mesozoic (cf. Jaekel 1915).

Ascidiidae. This large family of solitary Ascidians includes many species of one large genus on the continental shelf on the Atlantic side of the North American continent. These species are semitransparent, elongate and flattened zooids, some of which reach a large size, up to 90 mm. They have a large branchial sac without folds, but bearing internal curved papillae at the intersections of the transverse vessels. The loop of the digestive tube lies on the left side of the branchial sac. The gonads are within the loop and branch over the stomach and intestine. Small renal vesicles are found in the tissues around the gut.

One of the commonest species in the Gulf of Maine is *Ascidia prunum*, shown in Figures 15, 32a & h. It is gray and rather opaque. The dorsal oral siphon is well separated from the nearly lateral atrial opening, and neither disturbs the oval form of the zooid. When the test is excised, the zooid consists of a large branchial sac with the loop of the digestive tube on its left side. When a small piece of the branchial sac is removed and examined with a higher lens, the longitudinal and transverse vessels form square patches. At each of the corners, elevated rather massive papillae are seen, and there is one rather smaller intermediate papilla between. There are ordinarily four stigmata in each square.

The intestinal tract is large and massive, covering most of the branchial sac. The stomach is like an elongated pear and it tapers into the intestine. There are small renal vesicles in the tissues around the stomach. The gonads, especially the ovaries, lie within the intestinal loop close to the branchial sac. The finer extensions of the testis spread over the intestine. Rather small eggs (0.2 mm) are shed from the oviduct. They develop in the sea water and form small tadpole larvae which seek out favorable spots on the bottom at which they attach.

This species occurs in moderate depths from 20 to 100 meters and occasionally in deeper water. It appears to be circumpolar and is abundant in northern waters as it is north of Europe.

Another species is quite similar to the previous one, but it is found more commonly in somewhat shallower water, even close to shore in the Gulf of Maine, but

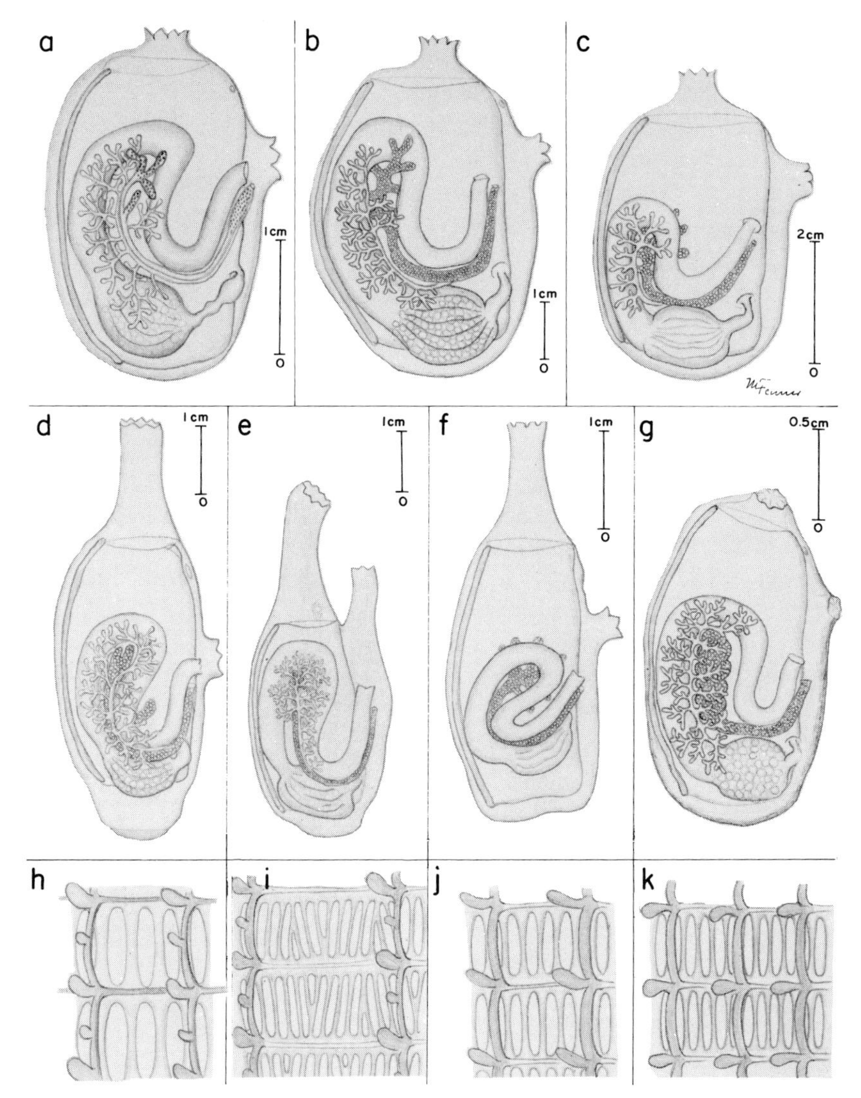

Fig. 32a–k. Species of *Ascidia* (after Van Name), with branchial sac detail, h–k. (a) *A. prunum*; (b) *A. callosa*; (c) *A. obliqua*; (d) *A. interrupta*; (e) *A. nigra*; (f) *A. curvata*; (g) *A. corelloides*; (h) *A. prunum*; (i) *A. callosa*; (j) *A. obliqua*; (k) *A. nigra.*

usually at depths of 25 to 150 meters. It is *Ascidia callosa* (Figs. 32b & i). In structure this species is much like the previous one. It usually has a firmer and thicker test, which makes these zooids a little more rounded and elevated. In favorable situations it may be even larger than *A. prunum*, with rather more prominent siphons.

There is a clear difference in the branchial sac, so identification requires dissection and microscopic examination of an excised piece of the sac itself. There are fewer longitudinal vessels than in *A. prunum*, so that the meshes of the sac of *A. callosa* are much wider, with fifteen or more stigmata in each. The two papillae between each transverse vessel are similar in each species, as are the gonads.

In sexually mature specimens, *A. callosa* can be easily distinguished when the test is opened by the presence of eggs and small tadpole larvae in the atrial cavity. This species is viviparous. The tadpoles are small but complete with sensory vesicle containing both a light sensitive organelle and a statolith.

The third species of *Ascidia* in the Gulf of Maine is *Ascidia obliqua.* This grows to a somewhat larger size than either of the other two and often has a softer, rougher test. Fresh specimens are often greenish or brownish. The siphons may be more extended or elevated than the others. When the test is opened and the upper half removed, it is found that the digestive tube is much smaller than in the others, apparently covering only about half of the branchial sac. The gonads have the same position within the loop. Like *A. prunum*, this species *A. obliqua* is oviparous.

Microscopic comparisons of a small portion of the branchial sac of *A. obliqua* with the other species easily distinguishes it. Dorsally curved papillae are found only at the corners of the meshes between transverse and longitudinal vessels. There are commonly five stigmata in each mesh (cf. Figs. 32c & j).

The distribution of *A. obliqua* in the Gulf of Maine is indistinguishable from that of *A. prunum*, but it appears to be less common. Like the latter it is found on Browns Bank and on Georges Bank, but seldom in the deeper waters between. Of the three species found in northeastern American waters, *Ascidia prunum* is the most successful in both moderate depths and deeper ocean areas.

The three species of *Ascidia* discussed up to this point (*A. prunum*, *A. obliqua*, *A. callosa*) are cold water forms found in deeper areas of the continental shelf north of Europe, Asia, and North America. All three are common specimens in any extensive dredging sampling from the Gulf of Maine, as shown by Marine Fisheries records for *Ascidia callosa*.

There are a number of *Ascidia* species found on the Atlantic continental shelf of North America south of the cold water areas in the northeast. The one which is best known and most widely distributed is *Ascidia nigra*, with a smooth gray-black colored tunic (Figs. 32e & k). These elongate zooids, with separated siphons bent close together, may reach a length of 70 mm and are attached to the bottom below and on the left side. It is found at depths of 10 meters or less off Sapelo Island, Georgia, off both Florida coasts, and quite commonly in West Indian waters. The structure of the branchial sac is similar to that of the northern *Ascidia obliqua*. The distribution of *Ascidia nigra* suggests that it may have reached the North American continental shelf via warm seas from Arabian waters.

Two additional species have a distribution suggesting similar geographic movement. They are *Ascidia interrupta* and *Ascidia curvata* (Figs. 32d & f). The first of these is an elongate sand-covered species distributed south from Cape Hatteras around both Florida coasts and throughout the West Indies. The second is smaller, with a smooth surface and often a pink or yellow color. It has been found at the Tortugas Islands and off Padre Island, Texas, by Nancy Rabalais of Texas Agricultural and Industrial University, Kingsville, Texas.

So far the *Ascidia* species described come from the Gulf of Maine or from the southern, West Indian regions. In our study of specimens collected by the *Albatross* from 50 meters depth south of the Bay of Fundy, one unexpected small *Ascidia*-like specimen was found. This is a small zooid, 12 mm long which was mounted because it looked like a *Rhodosoma* already described. Further microscopic examination suggests that it may be an *Ascidia* species like *Ascidia corelloides* (Fig. 32g). This is a typical *Ascidia*, but the intestine lies on the right side of the branchial sac. The gonads are quite similar to those already described for the new species of *Rhodosoma wigleii*, spec. nov. It is even more unexpected to find this small Ascidia-like specimen occurring in the northern part of the Gulf of Maine. The nearest recorded specimens, taken by Van der Horst in Caracas Bay and by W. Beebe in Prince Bay, Haiti, are clearly West Indian in distribution. If further collecting proves that this little specimen has to be given a new species name, its presence in northeastern waters suggests movement of some Ascidian species has taken place from south to north.

It was in his studies of the developing eggs of the European species of this genus, *Ascidia mentula*, that Kowalewsky (1867) found tadpole larvae and noticed their obvious Chordate affinities.

Pleurogona—Stolidobranchia

This is a large suborder containing most of the more successful solitary Ascidians. The atrial opening is formed from one median dorsal invagination, spreading within on both sides to form the atrial cavity. The branchial sac contains folds. The gonads are lateral and paired. Buds, when present, are lateral and from the mantle.

Pyuridae is a family of solitary forms sometimes attached by stalks. The test is rather leathery and often covered with spicules and spinous processes. The brancial sac usually has six folds. The stigmata are commonly straight, but are spiral in one family. Oral tentacles are usually branched.

The intestine forms a large loop on the left side of the branchial sac. The stomach has a large mass of hepatic tubules attached on one side. The gonads are elongate masses placed within the intestinal loop on the left and free on the right. Most species are oviparous, with the eggs forming complete tadpole larvae in the sea water.

The family is now divided into two unequal subfamilies following the studies of C. Monniot (1965). These are *Bolteniinae* and *Pyurinae*. It seems justifiable to add a third subfamily, the *Heterostigminae*.

Bolteniinae includes several genera widely distributed in northern boreal seas. One is *Boltenia ovifera*, which occurs north of North America, along the northern arctic to the Pacific coast, off Greenland, in Hudson

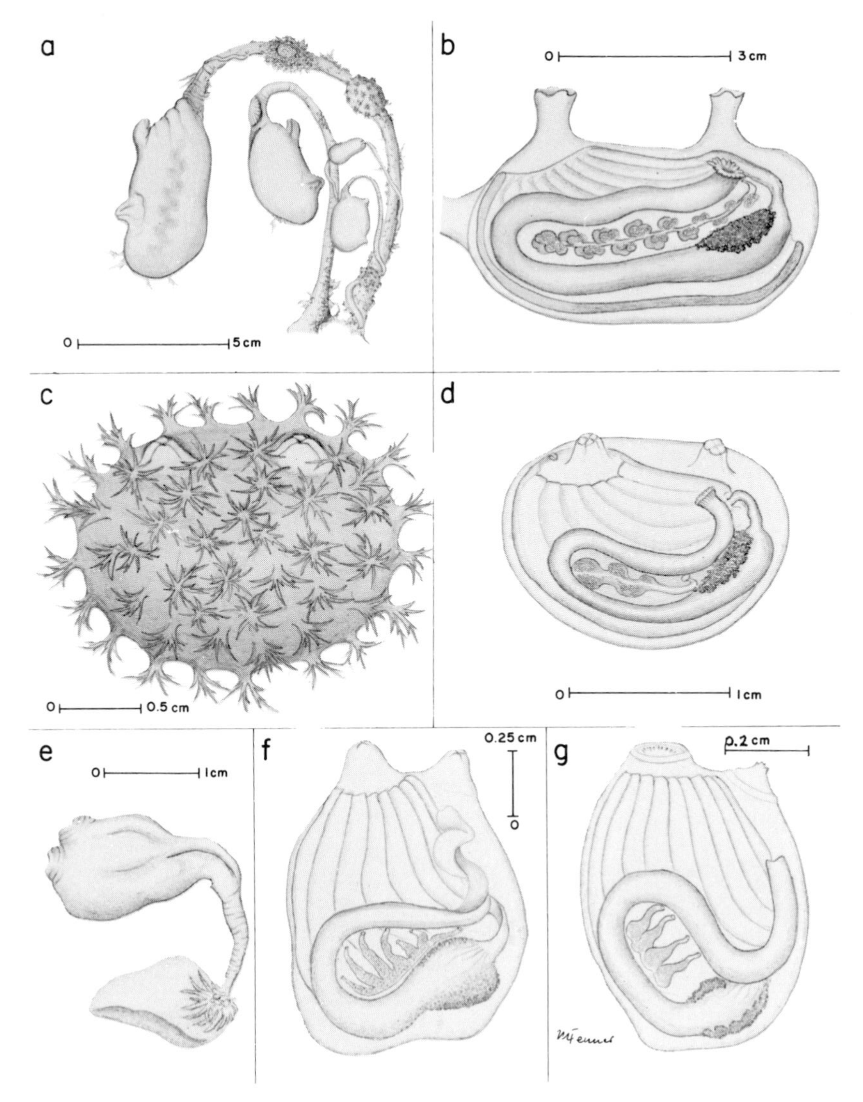

Fig. 33a–g. (a) *Boltenia ovifera* with young; (b) same, test removed; (c) *B. echinata*; (d) same, test removed; (e) *Halocynthia pyriformis*, young; (f–g) same, mature specimens, test removed.

Bay, and south into the Gulf of Maine to Cape Ann and Nantucket Shoals, on the northeastern Atlantic continental shelf of the United States (Figs. 33a & b).

Boltenia ovifera is an oval zooid 60 mm long attached by a stiff stalk at least 100 mm in length to stones or gravel at depths between 25 to 200 meters in the Gulf of Maine. The test is often colored, purple-red or pink-and-white, with the openings on the inverted dorsal side. As attached by one end to the stalk, it is tilted downward so that the atrial siphon lies below the oral. The species was first named by Linnaeus in 1767 and his name is still in use. The animals prefer to attach to a stony bottom on a site washed by the cold arctic current. The appearance of the live zooids is shown in the frontispiece, in color as they were brought up to the surface by the dredge of the collecting ship, U.S. Fisheries *Albatross IV* off Penobscot Bay on August 19, 1968. Many young zooids 10 mm to 20 mm were found attached to the stalks of the adult specimens.

In a recent paper (Plough 1969), it was suggested that since colored and white specimens, as well as some with thin transparent tests, were found at the same site, it is possible that there has been a mixture of genetically different stocks brought into the Gulf of Maine in different waves of migration from the north (cf. Frontispiece). In the northern sites in the Gulf as well as outside the entrance, *Boltenia ovifera* specimens are mostly of one color.

When the spicule-covered tunic is opened, it is seen that there are ten or more branchial oral tentacles. The dorsal tubercle is an inrolled C, and the dorsal lamina is cut into narrow teeth. There are nine or ten well-developed folds in the branchial sac, with many transverse vessels. The most striking feature is that the stigmata are placed with their long axis transverse to the folds (cf. Fig. 22f). This unusual arrangement suggests the first step in the formation of spiral stigmata which never appear. The digestive loop lies on the left side of the sac. The stomach tapers into the intestine. Together they form an elongate V. Several large greenish hepatic glands are found in the middle of the stomach facing the branchial sac.

The elongate gonads lie within the loop on the left side and alone on the right. The ovaries are within and partly surrounded by many small, rounded, testicular masses. The eggs are about 0.16 mm in diameter when mature in June. Tadpole larvae are so seldom found that it seems probable that eggs are shed infrequently, under physiological stimulus from the atrial siphon, and develop rapidly to the larval stage in the sea (Child 1922). The tadpoles often attach to the stalks or pedicle of the parent.

Boltenia echinata is a boreal circumpolar species which does not form a stalk. It is found over an even wider range than *B. ovifera*. When both are present, *B. echinata* may be attached to the stalk of *B. ovifera* or to nearby stones or shell (cf. Plate IX & Figs. 33c & d). It has no pedicle and is attached by its ventral surface with the slit-like siphon openings on top. Its striking appearance makes it one of the most unusual Ascidian species found on the northeastern continental shelf. The 10 to 20 mm long zooid has a flesh-colored or brown-red surface, which is covered by a cactus-like growth of large flexible spinous processes, each of which divides into six to eight bent, tapering branches. At first sight it resembles a sea urchin. Internally this species is similar to the larger species. Its branchial sac shows only six to eight folds. There is a line of longitudinal stigmata in the middle of each fold between the transverse stigmata which otherwise are arranged as in *B. ovifera*.

Boltenia echinata is normally viviparous and each specimen when opened in May or June contains many tadpole larvae in the atrial cavity. These are typical tadpoles, with two sense organs like those found in the *Aplousobranchia*. From the atrial cavity they are released into the sea water. When released from the living zooid they swim in the sea water for an hour or two before settling and metamorphosing. This is the most advanced Ascidian forming a complete tadpole from the egg.

A second genus of the *Bolteniinae* is *Halocynthia*. It is found off the Norwegian coast, Iceland, and across Davis Strait south to North America in the Gulf of Maine. Like the *Boltenia* species, its heavy tunic is covered with spinous papillae, but in *Halocynthia* these are small.

Halocynthia pyriformis, often called the "sea peach," is a heavy-bodied, often brightly colored zooid, attached at the base. In the Gulf of Maine it is dredged from hard bottoms at depths of 50 to 100 meters, but off Eastport it can be taken close to shore in less than 5 meters. Zooids of *Halocynthia* are among the largest of Ascidians of the northeast, reaching over 100 mm in length. They appear bright orange-yellow, light peach, or occasionally light scarlet. The siphons as crossed slits are often salmon color. The oral tentacles are branched.

The branchial sac has seven or eight folds on each side, with many longitudinal vessels. The stomach is short and broad with glandular folds near the esophagus and large hepatic glands on the upper side. The intestinal loop is wide and large. The tubular gonads are inside the loop of the intestine on the left side and free on the right. There are four to six massive elongate ovaries with testis massed at lower ends. Oviducts from each ovary unite into one, which opens into the atrial cavity. They are accompanied by sperm ducts (cf. Figs. 33f & g).

The eggs are released into the atrium, where a few at least are fertilized from outside, since ordinarily the male gonads mature at a later time than the eggs. Complete tadpoles with two sense organs have been found in the atrium (cf. Fig. 49b). These are released and swim briefly in the sea near the parental zooid, where they attach on the bottom or on stones or shell and metamorphose. The young zooids may show a different form from the parents. Dr. Tris Morse of Northeastern University found a number of specimens at Eastport up to 30 mm in length, which are elongated with the lower half forming a narrow pedicle as shown in Figure 33e. These apparently gradually enlarge to the typical adult *Halocynthia*.

Pyurinae. The second subfamily, includes several genera and species distributed off the coasts of South America, in the antarctic, and a few in the Pacific Ocean. Only two species have been dredged on the Atlantic continental shelf in our survey. *Pyura* is easily separated by the arrangement of gonads as separate sacs on either side of the oviduct. *Microcosmus* has one gonad passing partly through the loop of the intestine.

Pyura vittata is a species found in shallow water on sandy bottoms from the region of Beaufort, North Carolina, off Sapelo Island, Georgia, south around both Florida coasts, at Tortugas Island, and widely off most of the West Indian Islands. It is also found at Bermuda and as far south as the coast of Colombia. Its range over the world is even wider, since it is found off Hawaii, the Philippines, and the Japanese Islands.

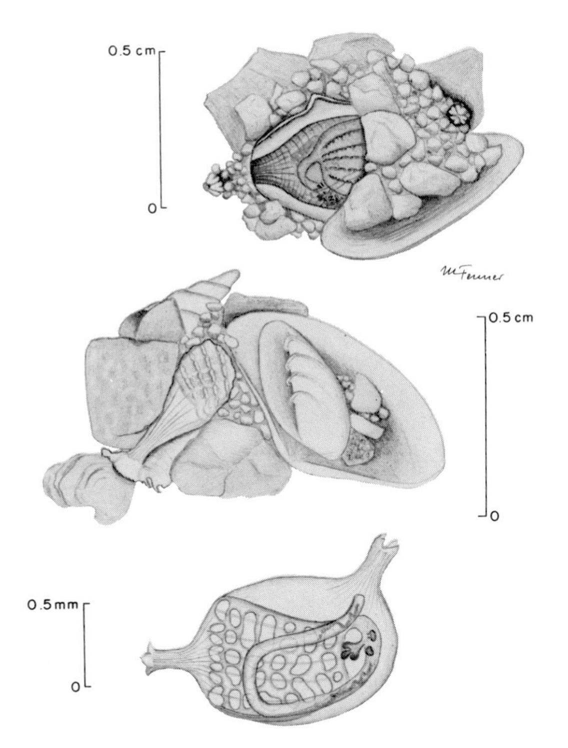

Fig. 34a. *Pyura vittata.* Young zooids in shell fragments, as found by Scott Leiper; exposed specimen.

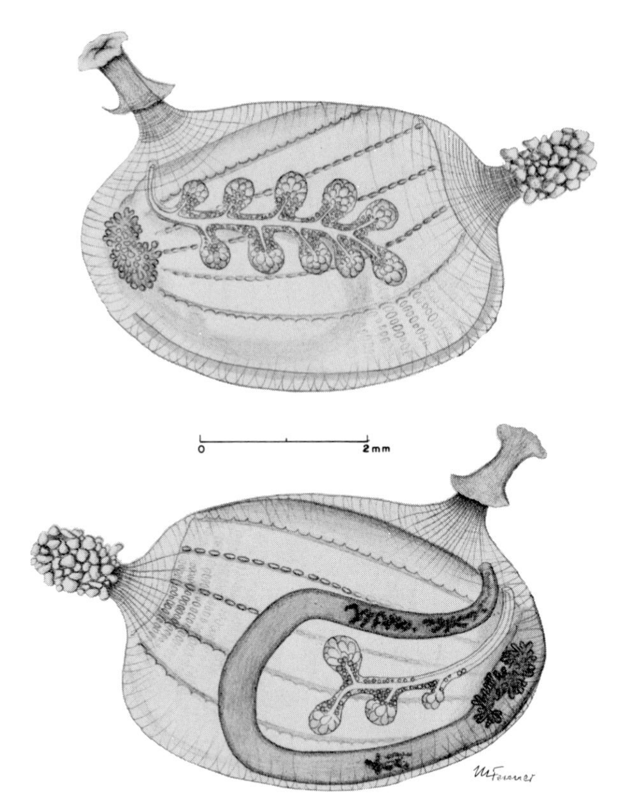

Fig. 34b. *Pyura vittata.* Young adult, right and left.

Our survey yielded specimens about two miles off shore from Sapelo Island, Georgia, from a depth of 25 meters. In the summer of 1970 young specimens—4 mm—were found half a mile off shore attached to shell or sand grains by Scott Leiper, University of Georgia (Plate IX & Fig. 34a). These specimens show that eggs and young are growing as interstitial animals quite close to shore along the Georgia coast. His work shows that accurate distribution of many common species can be given only by sampling at different depths off shore. The only location where our survey had shown *Pyura vittata* close to shore previously was at Tortugas Island, Florida (Fig. 34b).

This species is usually a rough, sand-covered zooid about 25 mm in length. Specimens appear gray or reddish brown, and the siphons are wide apart on the dorsal side. The branchial sac shows six folds. The most conspicuous characters are the double sets of gonads attached symmetrically along either side of the oviduct (see Fig. $34d_{1-2}$). It seems clear that *Pyura vittata* is an important suspension feeder a mile or two off shore in southern waters.

Another species of the *Pyurinae* is found infrequently along the southern part of the North American continental shelf. *Microcosmus* has been taken off Sapelo Island, Georgia, from 4 miles off shore at the region around the "Whistle Buoy." It is smaller than *Pyura vittata*, with similar external form. It is gray or purple-red in color. When a specimen is opened, however, differences immediately appear. This species shows nine or ten branchial folds, and the gonads are in five puffed masses. On the left side, the gonad passes inside the intestinal loop (Fig. $34c_{1-2}$).

This species is another Pyurid which is widely distributed in warm seas over the world. It occurs off Hawaii, Australia, and the Red Sea. Because old specimens are difficult to classify, it is rightly called *Microcosmus exasperatus.*

Heterostigminae. (New subfamily) Representatives of this subfamily are found north of Europe and along the coast of Sweden to the deeper water near the Azores. One species only is found south of Cape Cod in North American waters. *Heterostigma* has been studied by Ärnbäck–Christie–Linde (1924) and extensively by C. and F. Monniot (1961). Following Van Name (1945) all of these ascidiologists recommend placing these species in a separate subfamily, but this has not been done to this date. Kott (1969) separates them under the heading "Family?".

It is my recommendation that this situation be corrected by making a new third subfamily under the *Pyuridae: Heterostigminae.*

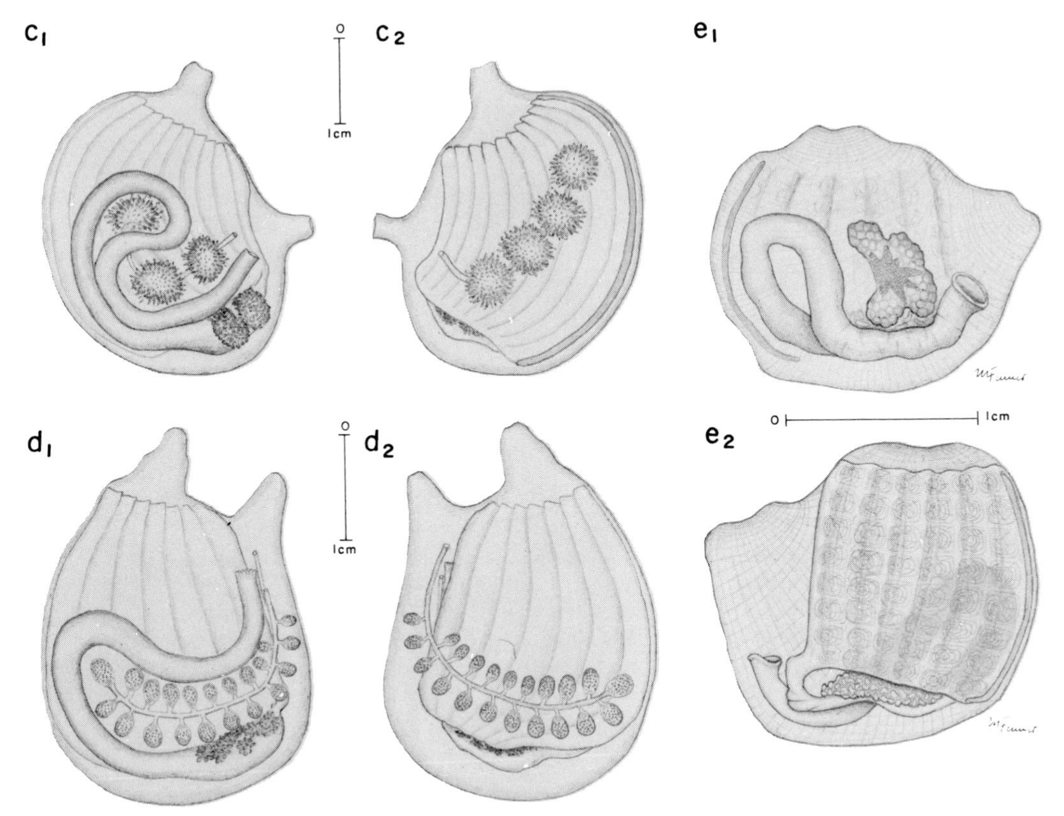

Fig. 34c–e. (c_{1-2}) *Microcosmus exasperatus*, right and left; (d_{1-2}) *Pyura vittata* (after Van Name), right and left; (e_{1-2}) *Cratostigma singulare*.

This subfamily is characterized as follows:

1) Branchial folds almost completely reduced, with their position indicated only by internal longitudinal vessels.
2) Symmetrical spiral stigmata in at least a part of the branchial sac.
3) Unbranched oral tentacles.
4) One hemaphroditic gonad on each side.
5) Elongate stomach with digestive glands, but renal organ absent.

The only species of this subfamily found on the Atlantic continental shelf of North America is the interesting *Cratostigma singulare*. This was first described by Van Name (1912) from specimens collected off Race Point and north of Block Island and named originally *Caesiria singularis*. Later, at the suggestion of Ärnbäck–Christie–Linde, he recognized its affinities with the Pyurids she had described and changed its name to *Heterostigma singulare*. The attempt of Van Name to give the correct Latin ending of the species name adjective with a neuter noun has been frequently disregarded.

Recently, C. and F. Monniot (1961) of the Histoire Naturelle Museum in Paris have pointed out that the species as described by Ärnbäck–Christie–Linde has spiral stigmata in the anterior part of the branchial sac and transverse slits of great length in the posterior part. Since Van Name showed that this species has spiral stigmata throughout the length of the branchial sac, it cannot be called *Heterostigma*, as was Ärnbäck's original species. Therefore, they made a new genus, *Cratostigma*, for one or two species found off the coast of Sweden, and also for Van Name's species. Thus, this species found close to the New England coast becomes now *Cratostigma singulare*.

So this little Ascidian with the frequently changed name once again receives attention. This species has seldom been recognized in collections of specimens from the northeast, mainly because it appears so similar externally to local *Molgula* and *Cnemidocarpa* species, yet it is entirely different when dissected. For some years specimens have been available in the preserved but unclassified material in the collections of the U.S. Northern Marine Fisheries Laboratory at Woods Hole from off Chatham, where it was identified by this author. I have since collected it in several other locations on Crab Ledge and other points close to Pollock Rip, and it has also been collected by Marine Biological Laboratory surveys off Chatham. Later the species was

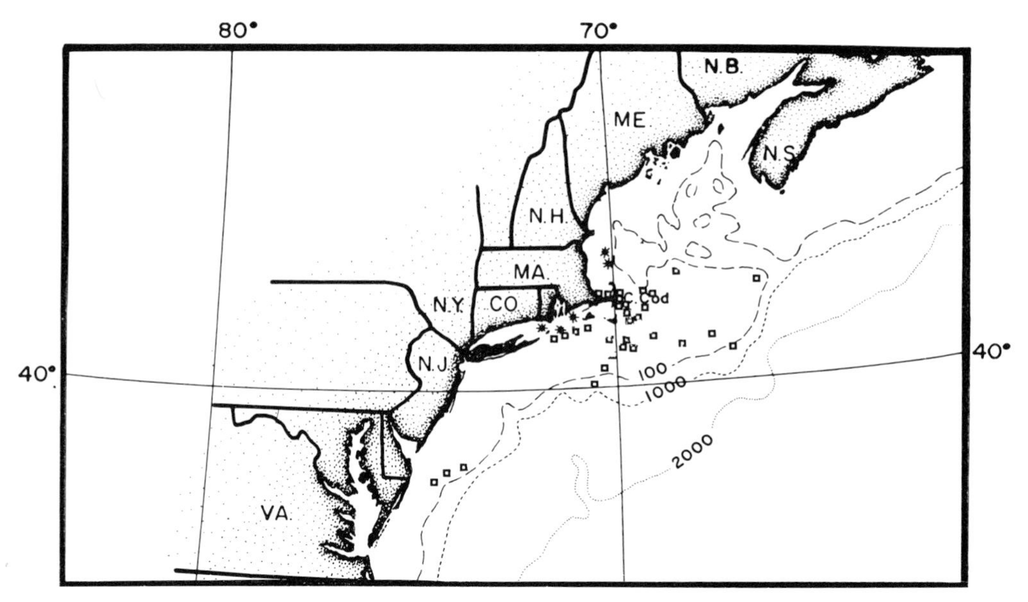

Fig. 35. Spread of *Cratostigma (Heterostigma) singulare*. Specimens collected to 1920 (asterisks). Specimens collected to 1970 (open squares).

found in deeper water off Pollock Rip in dredged specimens of the *Albatross*. Dredged specimens brought up on the WHOI survey of the *Gosnold* placed it at 40 meters in depth 5 miles off Cape May, New Jersey, and it has also been found in deep water near Georges Bank. Small specimens have also been identified in collections from sand close to shore off Barnstable in Cape Cod Bay. Thus, the species seems to be spreading as shown in dredging records (Fig. 35).

At the locations given off the southern angle of Cape Cod at Crab Ledge, I have found that *Cratostigma singulare* is frequently closely associated in attached masses with the Styelid *Cnemidocarpa mollis* and the Molgulid *Bostrichobranchus pilularis*. It seems clear that this little Ascidian, with the assistance of other sand-holding Ascidian species, is actually increasing its distribution in the Cape Cod area.

Cratostigma singulare is an elongate rounded zooid completely covered by fine sand which adheres to the test. It averages 10 mm in length and often has a pinkish shade due to the pink gonads. The mantle is thin with a rectangular mesh-like network. There are six low folds formed mainly by longitudinal vessels which are reduced in fold two. There are a series of spirally concentric stigmata which build up as flat infundibula to a single circle under the longitudinal vessels (see Plate X & Fig. $34e_{1-2}$).

The intestinal loop is very slightly bent and the anus has a widened opening. The narrow stomach has many hepatic papillae. The gonads lie as a crescent in the wide cavity of the intestine on the left side and are free on the right. The ovary is seen at the base of the three-lobed testes. In August, free eggs have been seen passing into the atrial cavity, but none have been seen developing. Probably eggs are viable at a certain time of day only. Frequent small pink crystals are attached inside the atrial chamber. They appear to be excretory products.

This is the species which Harant believed he had found in deep water south of San Miguel in the Azores. Possibly it has migrated across the Atlantic from Cape Cod, although the direction of the Gulf Stream currents are the only basis for such a migratory course. It seems too tiny and too dependent on attachment to fine sand, as at Cape May and Pollock Rip, to have migrated directly. There is no evidence of such migration by any other species. A more likely suggestion derives from the possible separation of Europe and North America in Mesozoic times. Perhaps the little zooids were present at the top of the inverted triangle when the split widened. Eventually, *Cratostigma singulare* was physically carried to widely separated areas within the separating land shelves.

C. Monniot (1965) places *Cratostigma* close to the trunk of the evolutionary tree from which *Boltenia* and most other Pyurids were derived. The same root may have given rise to the *Molgulidae*.

Molgulidae. This large family of solitary Ascidians has a simple structural plan which is little modified in the various environmental situations where it is distributed in the shallower portions of the Atlantic continental shelf of North America. Most of the many species are included within the single genus *Molgula*, which has representatives in shore in tidal locations, farther out on off-shore sandy areas, or in current-bathed deeper water favorable to masses of many different species. The most numerous species are tightly fastened to rocks, gravel, pilings or algae. A few are loosely attached in gravel or sand, and still others have tiny filaments on the base of the test holding them upright. Molgulids appear to have been distributed from the arctic southward, but, if so, more have become permanently established in European waters than over the shelf of North America.

The family includes many species with branchial folds and some in which these are secondarily lost. The stigmata are all spiralized by elongation and bending of the original straight slits. This greatly elongates the cilia-lined stigmata through which water passes into the atrium. The oral apertures have six lobes and the atrial four. The intestinal loop often extends dorsally, covering much of the left side of the branchial sac. A lobed digestive gland is present, attached asymmetrically at the base of the stomach. The most distinctive organ is the closed renal vesicle or sac-like "kidney" on the right side of the body outside the branchial sac. The gonads usually lie in the secondary loop of the intestine on the left side and above the renal sac on the right. *Molgula citrina* and *Molgula complanata* are viviparous, but in most species the eggs are shed into the atrial cavity and develop into tadpole larvae after passing out into the sea water. In several species in this family the tadpole is reduced by loss of the tail, or no tadpole at all is initiated from the developing egg.

The *Molgulidae* must have evolved from one branch of the *Pyuridae* with a similar body plan (cf. C. Monniot 1969). Instead of a number of fine renal sacs, a large, single, renal vesicle is the major organ added. The *Molgulidae* are usually considered the most advanced of the Ascidian families. This is not necessarily true, and they show no such diverse branches as can be found among the *Styelidae*. The family is very ancient in geologic time, as shown by the wide distribution of the species. But its lack of diversity suggests it is not as ancient as the *Styelidae*. It is for these reasons that the *Molgulidae* are considered at this point in the review, and the *Styelidae* will be described along with some of the diverse subfamilies at the end of the systematic survey. An excellent survey of the *Molgulidae* of European seas is given by C. Monniot (1969).

Molgulidae are divided into two unequal subfamilies on the presence or absence of branchial folds: *Molgulinae* with branchial folds, commonly six or seven on each side; and *Eugyrinae* without branchial folds but with longitudinal vessels in the position of folds.

Molgulinae of American waters may be arbitrarily separated into four groups on the basis of their branchial sac diversities, their distribution, and the extent of the degeneration of their tadpole larvae.

1. The first group includes the two viviparous species, *Molgula citrina* and *Molgula complanata.* Both species are found in certain shallow harbor areas where there is an active current and also on sandy bottoms with some gravel, as in Vineyard Haven at Martha's Vineyard or in Cape Cod Bay and Cape Cod Canal. Each of them sheds mature swimming tadpoles, bearing only one head sense organ, which seek bottom attachment after swimming to the surface (Grave 1926).

Molgula citrina is a rather small species (20 mm) with rigid apertures well separated (Plate XI & Fig. 36d). It is usually attached by a broad ventral area. When the test is removed it is seen that the intestine is a flattened loop and the kidney is large and flat. The gonads form S-shaped masses and the long, slender oviducts bend dorsally. The branchial sac has seven narrow folds with heavy longitudinal vessels above them. The stigmata are straight, bent spirally at the corners, and form a series of doubled infundibula under the vessels at the summits.

Molgula complanata is a small species, usually 5 to 10 mm long, which has been called by many names (Plate XI & Fig. 36e). The siphons are pink-red in life, and specimens are found which are all red. Zooids are commonly attached to gravel or stones in waters of moderate depth by slender fibers from the test. We have found it in Cape Cod Bay and from Georges Bank. In 1967, D. Frame found it to be common in the Cape Cod Canal at 5 meter depths. The siphons are close together and are extended at least half the length of the body. The common, rounded form of massed zooids attached by the ventral side sometimes changes to flattened by more complete attachment laterally. Removal of the tunic always shows large mature tadpoles crowded in the atrial cavity during July and August. The tadpoles swim about thirty minutes to an hour.

There are usually six branchial folds with regular spiralized stigmata forming infundibula under the longitudinal vessels. The intestine shows a flattened loop. The dorsal lamina is a narrow straight fold. There are many short separate hepatic tubules as the esophagus passes into the stomach. The renal vesicle is on the right side, flat, and wide. On each side, the gonad ducts are unusual. They are bent into an S shape, with the short oviduct often pointing downward away from the atrial siphon through which the eggs must pass. Thus the eggs are encouraged to remain in the atrium.

It is of considerable interest that several small specimens of *Molgula complanata* have been collected close to shore in protected crevices in Matagorda Bay in shore from Padre Island, Texas (cf. Fig. 37a). The

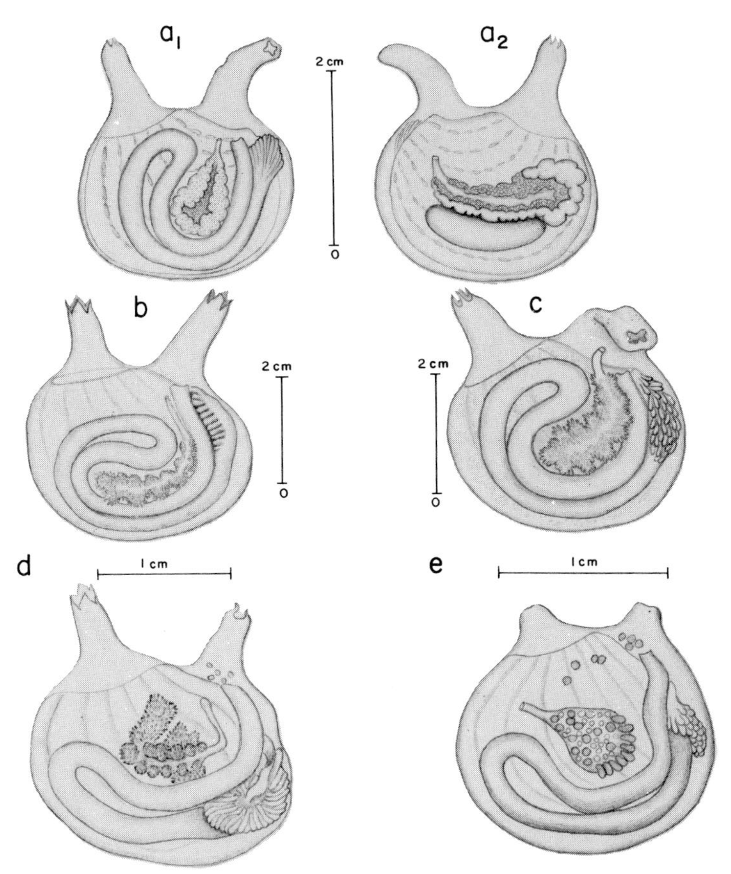

Fig. 36a–e. Species of *Molgula* (after Van Name). (a_{1-2}) *M. manhattensis*; (b) *M. provisionalis*; (c) *M. robusta*; (d) *M. citrina*; (e) *M. complanata.*

zooids and tadpoles are a dark red color which is not dissolved in alcohol. These specimens were found by Nancy Rabalais and associated collectors from Texas Agricultural & Industrial University in Kingsville, Texas.

The specimens are small, about 3 mm but are sexually mature. There are a number of mature tadpoles ready to swim in the atrial cavity. As shown in Figure 37b, these tadpoles already have developed the rudiments of stomach, intestine, and heart usually found only after several hours of swimming.

The location where these specimens were found is very much farther south than specimens of *Molgula complanata* have ever been reported before. Indeed, they have not been picked up south of Long Island Sound on the Atlantic continental shelf. The Texas Agricultural & Industrial University group deserves much credit for picking up such minute but interesting specimens from Matagorda Bay, Texas.

Each of these viviparous species is found in many locations off Europe and in the boreal subarctic (cf. C. Monniot 1965). As small viviparous species with a northern circumpolar distribution, they appear to be the most primitive of living *Molgulinae*, evolved from Pyurid sources and widely distributed in temperate zones in northern hemisphere. Although swimming tadpoles maintain these species even on sites with ocean currents, the tadpoles show some reduction compared with other families in the presence of one major sense organ, the statolith, rather than two, as in *Pyuridae.*

Among the remaining Molgulids still to be described, several show further reduction in tadpole structure. In many species the tadpole is very small and inactive, the tail is reduced, and in several *Mogulinae* and *Eugyrinae* the tail of the tadpole is no longer formed and the egg develops at once into the adult Ascidian. This sequence is one of the most interesting of evolutionary transitions, since a once important organ—the tadpole with its notochord and dorsal nerve cord, suggesting vertebrate affinity—is completely lost and disappears as a development stage (cf. *Molgula retortiformis*).

2. The second group includes the tidal in-shore species *Molgula manhattensis*, *M. robusta*, *M. provisionalis*,

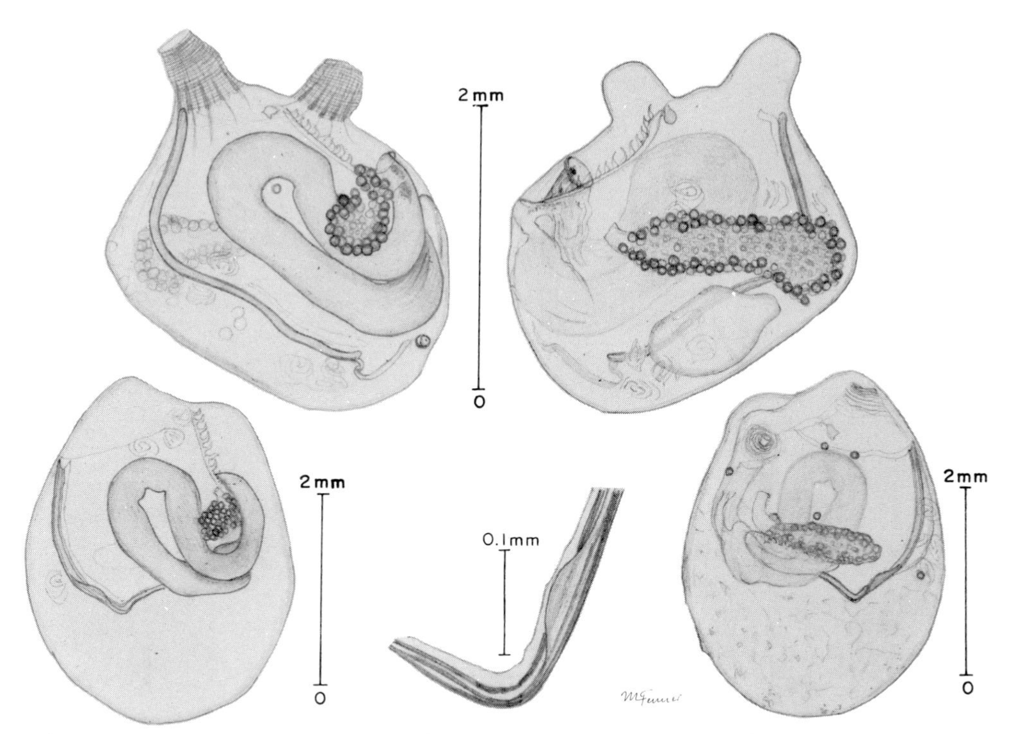

Fig. 37a. *Molgula complanata.* Zooid, right and left; another zooid, right and left, with enlargement of heart rudiment, lower center. Specimen collected by Nancy Rabalais from Matagorda Bay, inside Padre Island, Texas.

and the southern *Molgula occidentalis.* These are all shallow water species, growing from low-tide level to 20 meters depth, attached to stones, wharf piles, or seaweed. *M. manhattensis* is the commonest shore-growing Ascidian off the eastern coast of North America, but many others are just as common in deeper water on the continental shelf. It is a species which persists where the salinity is reduced in harbor areas by the emptying of rivers, and it appears to be relatively insensitive to pollution from the shore. Very few other Ascidian species are so insensitive to materials poured into the sea from human agencies.

In coastal ocean areas south of Cape Hatteras the ubiquitous *M. manhattensis* is superseded by the larger *Molgula occidentalis.* Several other less common species, or perhaps subspecies, are found in the northeastern coastal areas, especially along the Gulf of Maine and in Cape Cod Bay or Long Island Sound.

Molgula manhattensis has a globular form, averaging 25 mm in diameter, with a tough tunic bearing patches of short, hair-like papillae to which adhere sand grains, shell bits, or algal fragments. The living specimens often show a yellowish-green color. The siphons are close together but diverge and remain extended even when contracted (Plate XII & Fig. $36a_{1-2}$).

When the test is removed, the transparent zooid shows on the left side the characteristic circular tight loop with the gonad filling the second loop. On the right, the gonad is above and extends beyond the bean-shaped kidney. The dorsal tubercle is C-shaped and opens downward. The dorsal lamina is slightly toothed. The branchial sac has six folds on each side. In young individuals the spiral stigmata form large infundibula with the point of the funnel under the fold. This arrangement becomes irregular in older specimens.

The eggs are shed into the atrial cavity and from there are released through the atrial siphon into the sea water, where they develop into small swimming larvae. Grave (1926) found the eggs measured 0.11 mm and the tadpole larvae grew to 0.75 mm long.

Molgula robusta is found occasionally off Martha's Vineyard and Buzzards Bay (Fig. 36c). It is somewhat larger than *M. manhattensis*, reaching a length of 35 mm. The body is coated with sand and rests on the bottom (off Edgartown) without additional attachment. The siphons are not as extended and the intestinal loop is more open than in *M. manhattensis.* The branchial sac has usually four longitudinal vessels over each fold, and the stigmata are short and bent around into spirals only at the top. The reddish or purplish eggs develop directly into small adults without forming a tailed tadpole larva, unlike *Molgula manhattensis.*

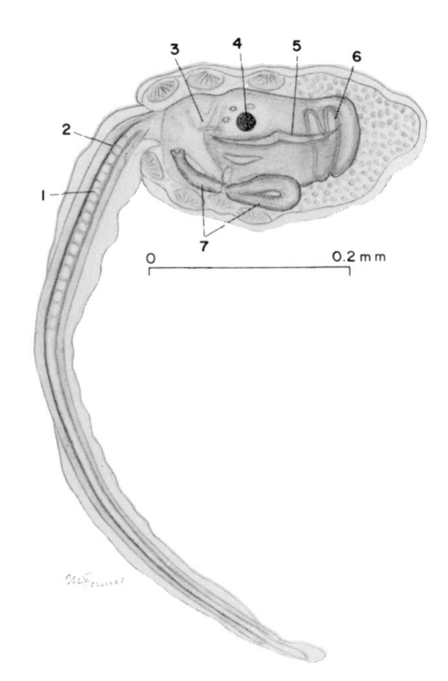

Fig. 37b. *Molgula complanata* tadpole. (1) notochord, (2) nerve cord, (3) atrial siphon, (4) otolith, (5) endostyle, (6) oral siphon, (7) alimentary canal.

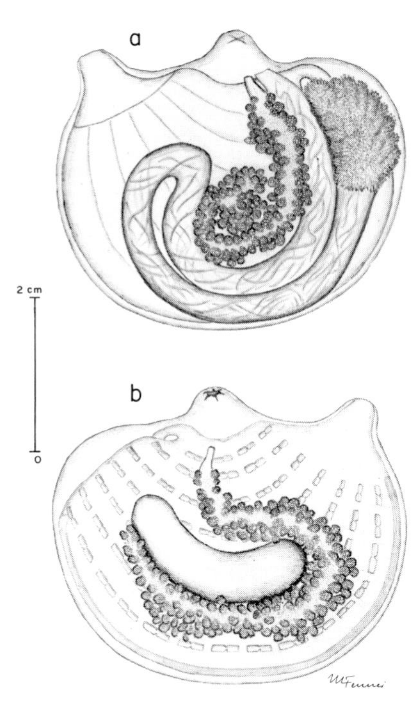

Fig. 38. *Molgula occidentalis* (a) right; (b) left.

Molgula provisionalis was described in 1945 to distinguish specimens found originally off St. Andrews, New Brunswick, and at the mouth of the Bay of Fundy (Fig. 36b). It seems to be present infrequently as far south as Vineyard Sound south of Cape Cod. The stigmata are arranged in more regular infundibula than in *M. manhattensis*. The chief differences are that the intestinal loop is more open and the gonads are more striated, with the male glands surrounding the ovaries. The eggs are larger than *M. manhattensis*—0.18 mm—and develop directly to the adult stage, like *M. robusta*, and like the European form *M. macrosiphonica* which this species resembles.

Finally, the much larger southern species, *Molgula occidentalis*, should be compared (Fig. 38). An occasional specimen is picked up north of Cape Hatteras, and it is numerous in the shore waters off Florida on both coasts. Specimens have been found off lower California in the Pacific Ocean and west of Italy in the Adriatic. This specimen has a leathery tunic and it reaches a length of 100 mm and 40 mm across. Its intestinal loop is more separated than *M. manhattensis*. Like the two previously mentioned species, the eggs of *M. occidentalis* have lost the ability to form swimming larvae and develop directly to the adult stage after being shed into the sea water.

3. A third group of species are found loosely attached to rather hard sandy or gravelly bottoms, usually farther out to sea. They are northern or subarctic species which have moved south: *Molgula siphonalis*, *Molgula retortiformis*, and *Molgula griffithsii*.

The first, *Molgula siphonalis*, is distributed from the Arctic and off Greenland to the seas off northern Europe and North America. On the American side it has been dredged from off Labrador, the Gulf of St. Lawrence, and around Nova Scotia into the Bay of Fundy. Off Eastport, Maine, and in Casco Bay it has been found in quite shallow water, 10 meters. Farther south it is dredged from quite widely scattered locations washed by the Labrador current on eastern Georges Bank and down to Cape Cod Bay. The elongate oval, rather flattened zooids are covered with coarse sand and are attached loosely to the bottom by hair-like processes from the test (Fig. 39a).

The siphons are rather close together and may be greatly extended. When the tunic is removed, two radiating muscle bands are seen to descend around each siphon, and two strong muscle bands pass along on each side of the endostyle. There are seven folds on each side of the branchial sac with a series of spiral stigmata forming two elevated infundibula under the folds. The intestinal loop is rather flattened and the left gonad lies above it. On the right side, the large kidney sac is surmounted by the right gonad. The species is oviparous, and the eggs develop in the sea water directly to the adult stage without the formation of a tadpole.

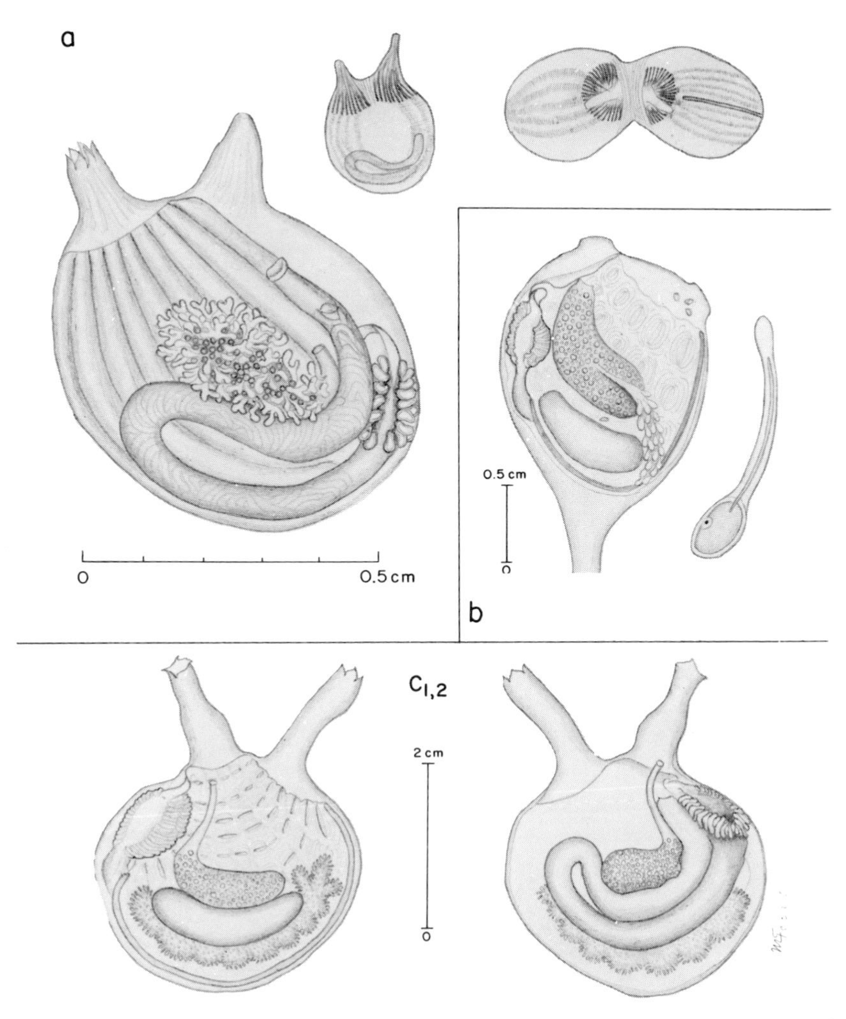

Fig. 39. (a) *Molgula siphonalis*; (b) *M. griffithsii*; (c_{1-2}) *M. retortiformis*, right and left.

Molgula retortiformis is a very large Ascidian (75 mm long) with a tough tunic which is distributed from the arctic even more widely than the previous one. It is found southward in European and North American Atlantic waters and also in the North Pacific. Like the previous species it has two very elongate siphons, the atrial often half the length of the body. The species is distinctive in that it is the only Ascidian with the male and female gonads separate. The female portion is above on each side. On the right, the female gonad is above the kidney and the male below. There are seven branchial folds, with four or five longitudinal vessels following each. The stigmata bend noticeably only at the top under the vessels where they form infundibula. In the gonad the male germ cells are formed at the upper edge. The eggs of this species are shed into the sea from the atrial cavity. There they develop directly to the adult stage without forming tadpoles (Plate XII & Fig. 39c_{1-2}).

The third species is less common, *Molgula griffithsii*. It is small, 10 mm or less. Like the previous species, these zooids are found at moderate depths far from shore. Unlike the two previous species, it is attached on sandy bottoms by a short pedicle or stalk, so that the atrial siphons are above the oral. There are only five branchial folds and the stigmata form regular spirals. In the gonads the male cells are well separated, and on the right may extend posteriorly around the end of the kidney. In this species, however, eggs form a typical tadpole larva during development to the adult stage (Fig. 39b).

4. Next we come to the fourth group of *Molgulinae* on the North Atlantic continental shelf. Included are: *Molgula arenata*, *Molgula lutulenta*, and perhaps the rare *Molgula habensis*. The first of these, *Molgula arenata*, rivals the shore-living *M. manhattensis* in its wide distribution on sandy bottoms at moderate depths from 2 to 20 miles off shore. It used to be described

as a temperate living species distributed from Cape Cod to Cape May, but in the past fifty years it has been found further south. It is now one of the most numerous Ascidians on the wide expanse of the continental shelf from Cape Hatteras to Florida. East of the University of Georgia Marine Institute at Sapelo Island, it can be brought up with the naturalist's dredge almost anywhere from 1 to 4 miles off shore at 15 to 30 meters depth, and the zooids will be 20 to 25 meters long. They appear to lie on the bottom without hold-fast organs, but they prefer positions where a slow south to north marine current is present. This off-shore Ascidian shares with the larger shore-dwelling *Molgula occidentalis* the position as the most numerous species in southern portions of the continental shelf.

Molgula arenata has widely separated siphons and a rather flattened intestinal loop. The branchial sac shows six folds, with regular spiral stigmata which are topped by two infundibula each. The gonads are large, with the male elements on the outside. The right gonad is lapped over the distal end of the kidney. Young specimens have been found closer to shore in the shell-fragment area. In these locations they do not reach a size greater than 10 mm. Farther out they reach a length of 10 to 20 mm. *Molgula arenata* is a common species associated with many others in the live habitat areas, often held in position by algal and other Ascidian body masses and closer to shore than larger numbers of the species are usually found (Plate XII & Figs. 40 & 41).

The other species in this group are similar in structure but adapted to deeper water off Florida and West Indian islands.

The *Eugyrinae* are a subfamily of the *Molgulidae* which differ considerably in structure from *Molgula.* They tend to favor protected pockets as muddy areas where there is a current above. They possess no branchial folds but they have widely spaced longitudinal vessels. Two narrow stigmata form a series of clear double spirals with two infundibula at the upper end under the longitudinal vessels. The kidney is on the right side, as in other Molgulids. There is only one gonad placed inside the primary loop of the intestine. These characteristics apply to the genus *Eugyra* found in Europe and on the pacific coast of North America.

The Atlantic continental shelf has a somewhat distinctive member of this subfamily given a distinct genus name: *Bostrichobranchus pilularis.* In this species the spirally coiled stigmata continue their growth, following down from the infundibulum summits to the areas in between. There each begins a secondary coil to form an intermediate infundibulum. This process continues until in older zooids the branchial sac becomes a tufted surface of indiscriminately directed funnels showing a nippled appearance. This structure appears to be a successful gill for the muddy or fine-sand environment which this animal prefers. It has a greatly enlarged length of stigmata or gill slit openings.

When sexually mature *Bostrichobranchus* shows numerous eggs within the atrial cavity, but no tadpoles are formed. The eggs develop directly to the adult structure and settle in the same area where the parent zooids are. Thus, here is another large subdivision of the Molgulid family in which the formation of notochord-bearing tadpole larvae has been discontinued (Plate XIII).

Bostrichobranchus pilularis is distributed off the whole length of the coast of the United States. It has been found in the St. Lawrence Estuary and off Newfoundland. Irregularly, it appears in shallow water within one-half mile of shore off the Maine coast. Often *Bostrichobranchus* appears in large numbers in live habitat areas south of Cape Cod (Cuttyhunk) and, less frequently, in similar areas farther south. It has been dredged off Sapelo Island, Georgia, at the "Sponge Reef," at two or more southerly stations on the Florida east coast, and off Lemon Bay on the west Florida coast. But it is predominantly a cool water species. It is in the live bottom areas that *Bostrichobranchus* zooids are largest, up to 25 mm or larger. Here they appear to have a life span of two years or more (see distribution chart, Fig. 5).

A related variety or subspecies has been collected off Padre Island southeast of Corpus Christi, Texas, by Nancy Rabalais of Texas Agricultural and Industrial University. This variant is smaller than the previous species, but it has nonretractable siphons which are heavily spirally lined with longitudinal stripes. Thus they resemble *Eugyra arenosa* of European coasts. These will be described later. Currently they are called *Eugyra arenosa padrensis.*

In a review such as has been given here of all the species of the family *Molgulidae* on the American Atlantic continental shelf, certain broad conclusions about the origin and evolution of the family are suggested. The presence of both *Molgula citrina* and *Molgula complanata* on the continental shelves of Europe and eastern North America fits very well the view that these species were placed in Europe and North America as these continents were progressively separated following the split and separation in Mesozoic times. These tadpole-forming species probably were present at the northern end of the cleft as the Atlantic Ocean formed in Mesozoic times, so each was present on European and North American coasts. The tadpole larva was of importance in the initial distribution into the cooler off-shore habitats. From this start one line went to the tidal areas. *M. manhattensis* was successfully distributed here with a reduced tadpole stage. Another went into deeper water following colder northern currents. *Molgula si-*

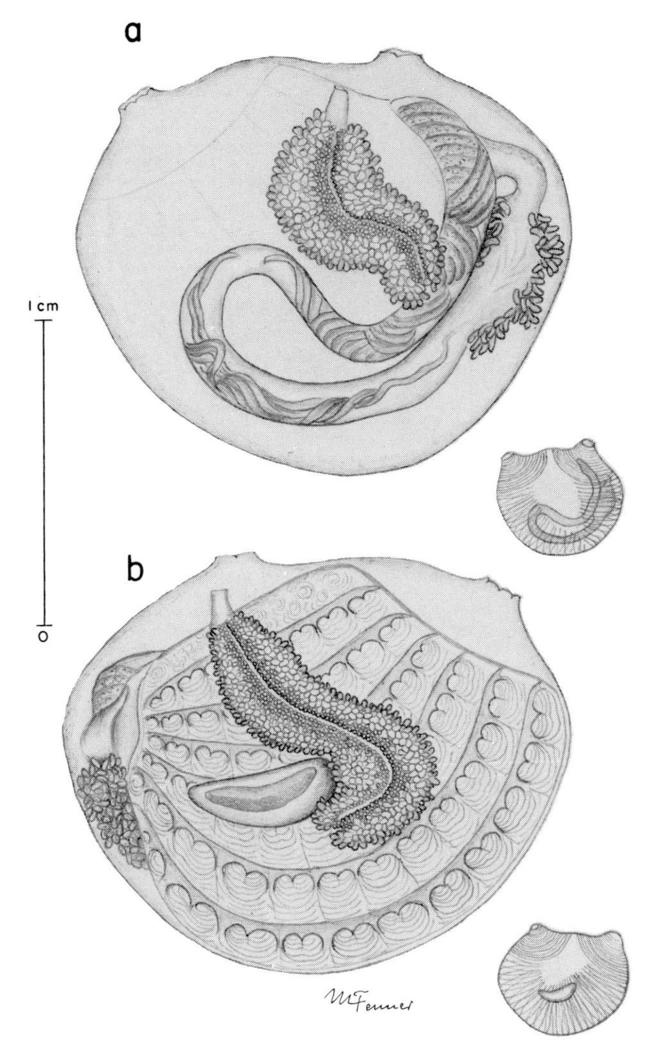

Fig. 40. *Molgula arenata.* (a) left; (b) right.

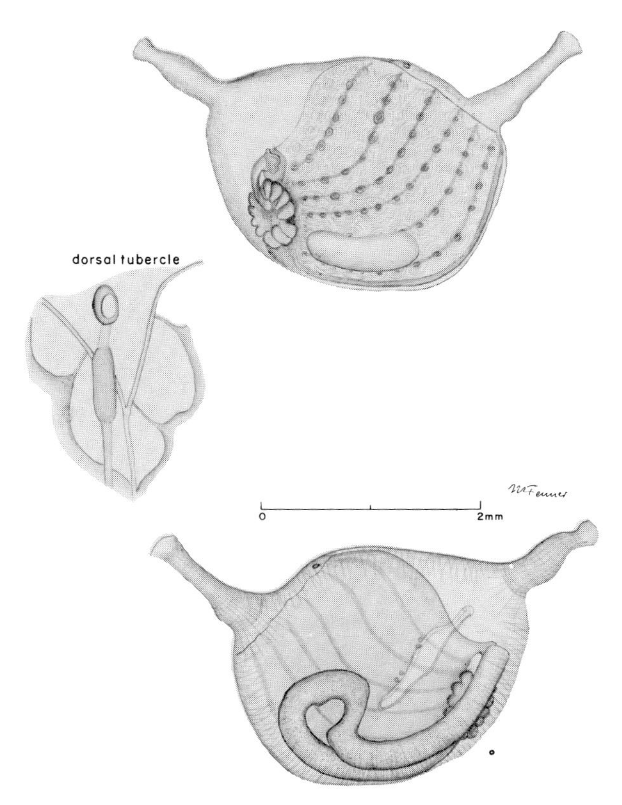

Fig. 41. *Molgula arenata.* Young zooids interstitial in sand off shore at Sapelo Island, Georgia, found by Scott Leiper. Right, left, and detail of dorsal tubercle.

phonalis and *M. retortiformis* continue to cultivate this habitat without tadpole larvae. *Bostrichobranchus* represents the *Eugyrinae* branch which has found in-shore pockets for mud dwellers. These also have found tadpole larvae unnecessary for their preservation. Thus, as far as the American Atlantic continental shelf Ascidians go, we are forced to assume that the ancestors of chordate and eventually vertebrate animals probably developed before most Molgulids appeared and probably from still more ancient predecessors. *Molgula citrina* could have given rise to a neotenous line of free-swimming tadpoles with mature germ cells, but the actual history in *Molgulidae* appears to have gone toward the loss of the tadpole, both for the in-shore dwellers, and with the dwellers in the smooth sand under the off-shore currents.

Styelidae. This is the largest and the most diversified of all the Ascidian families. The branchial sac commonly has four folds. The stomach has ridges and often a pyloric cecum. No one of the genera shows the unusual excretory reservoir of the Molgulids, nor the hepatic diverticula of the Pyurids. But within the large family many structural and physiological specialties appear, like the disc-like flattening of *Dendrodoa* on stony bottoms or the gonad and endocarp multiplication in *Polycarpa* on deeper sand and mud. Most Styelids, like *Polycarpa*, have a well-developed functional tadpole and the larva helps in maintaining the species in deeper waters. None of the *Stolidobranchia* concentrate vanadium, but the *Styelidae* utilize excess manganese which may give them greater adaptability than the other advanced families.

Styelid species are found in all the ocean environments from arctic to tropical and from close to shore to far out on the continental slope. One species, *Styela partita*, is among the most numerous of sea squirts collected from the shore. Another, *Styela nordenskjoldi*, found in deeper water off the shore all around the antarctic continent is possibly still more numerous but is seldom seen by men. They show such world-wide distribution that the *Styelidae* are considered as the most advanced and most successful of the families of the *Stolidobranchia* branch of the *Ascidiacea*.

In addition to many genera of the solitary *Styelidae*, there are also several widely distributed colonial species. For convenience, the colonial forms are arranged into two subfamilies: *Botryllinae* and *Polyzoinae*, and the large number of solitary genera are placed in the subfamily *Styelinae*.

All *Styelidae* have four lobed siphon openings and the branchial sac has no more than four folds. The stigmata are straight. The stomach wall is folded or ridged, and there is a functional, curved gastric cecum. The intestinal loop is found on the left side of the branchial sac. The large gonads are attached to the inner surface of the mantle on either side and are often visible through the mantle.

In the compound Styelid species the folds of the branchial sac are reduced or absent. The buds are formed as outgrowths from the lateral body wall, either paired or asymmetrically.

The eggs of *Styelidae* develop into functional tadpoles which differ from those of *Ascidiidae* or *Pyuridae* in having one functional sense organ, an otolith, with the ocellus reduced or lost. The loss of the eye spot is compensated for in *Botryllus* to some extent by some light sensitive cells within the otolith, so Grave called it a "photolith."

A solitary Styelid, *Polycarpa fibrosa*, of wide occurrence in the North Atlantic, shows the general plan of structure seen in this family, and its distribution suggests the mode of life which was that of the earliest Styelid species on the North American continental shelf (Plate XVI & Figs. 18 & 42). Farther south a similar position is taken by *Polycarpa obtecta* or *Polycarpa circumarata*. It is of interest that the habits and distribution of these species are rather similar to those of *Molgula siphonalis* and *Molgula arenata*, which may be the earliest North American Molgulids.

Polycarpa fibrosa is found off both coasts of Greenland. On the European side it occurs in the English Channel and as far south as Portugal. Off North America it ranges from the Gulf of St. Lawrence into the Gulf of Maine. It is found here in the colder waters covered by mud or fine sand at water depths from 50 to 300 meters. Also it is dredged from close to shore off Eastport and Mt. Desert Island south to Georges Bank and Nantucket Shoals. The extensible siphons reach up through the mud more than twice the thickness of the zooid.

Polycarpa shows a generalized Styelid structure. After removal of the test it is seen that there are four folds in the branchial sac, and the stigmata are large and straight. The dorsal tubercle is C-shaped and the dorsal lamina is a smooth membrane. The stomach has many fine folds and ends in a protruding cecum. The intestine is a wide loop. There are twelve gonads in the mantle on each side, together with many clear endocarp sacs. The gonads are sac-like with an internal ovary, and many externally extruded testis lobes. Oviducts and sperm ducts are separate. It appears that from such a generalized cold water, off-shore species the many specialized species of the Atlantic continental shelf were derived.

Two additional distinctive Styelid lines should be mentioned. One is a northern circumpolar form with a specimen in the National Museum which has been found off Sable Island (Van Name 1912–44). It was reported more than a hundred years ago from the Gulf of Maine in about 30 meters east of Boston. It is an elongated sac-like specimen reaching 100 mm, but usually much shorter. The branchial sac has no folds and it extends only two-thirds of the body. A long gonad is located on each side. This unusual species is called *Pelonaia corrugata* (see Fig. 44e). Its simplified character suggests *Ciona*, and it appears to be an early evolutionary experiment of the *Styelidae*, rather than a reduced tag-end of an ancient line.

A frequent modification of the standard Ascidian pattern of a mud-immersed sac with siphons extending upward into the sea water above (like *Polycarpa* or *Pelonaia*) is the flattened zooid attached by the left side on a rock elevated above the sand and with the opened siphons only slightly raised above the flattened surface (*Dendrodoa*). This adaptation has occurred independently in *Rhodosomatinae* (*Rhodosoma wigleii*), in *Corellinae* (*Chelyosoma macleayanum*), and in several small species in the family *Ascidiidea* (*Ascidia corelloides*), as well as in a number of species in the *Styelinae*, all in the genus *Dendrodoa*.

The species of *Dendrodoa* are mostly subarctic, common in marine locations bordering Greenland. At least two species are found southward in the Gulf of Maine as far as Cape Ann, and one at least in the waters off Block Island. The species *Dendrodoa pulchella* occurs in groups of three or four, all attached firmly by the ventral side. Its structure is much like *Pelonaia*, with reduced folds in the branchial sac, but with three gonads only on the right side. Although the three gonads all contain ovary and testis, they appear to be joined by their bases into one united organ. This is much the way it appears in the northern species, *Dendrodoa aggregata*, in which there are regularly more than three gonad lobes. *Dendrodoa pulchella* seems a distinctive species in the colder water close to shore off Cape Ann (see Figs. 20 & 44f & g).

A more common, flattened, tightly attached zooid found widely in the Gulf of Maine and through Vineyard Sound as far as Long Island Sound was set aside by Traustedt as a different subgenus, *Dendrodoa* (*Styelopsis*) *carnea*. It is regularly flattened dorsoventrally,

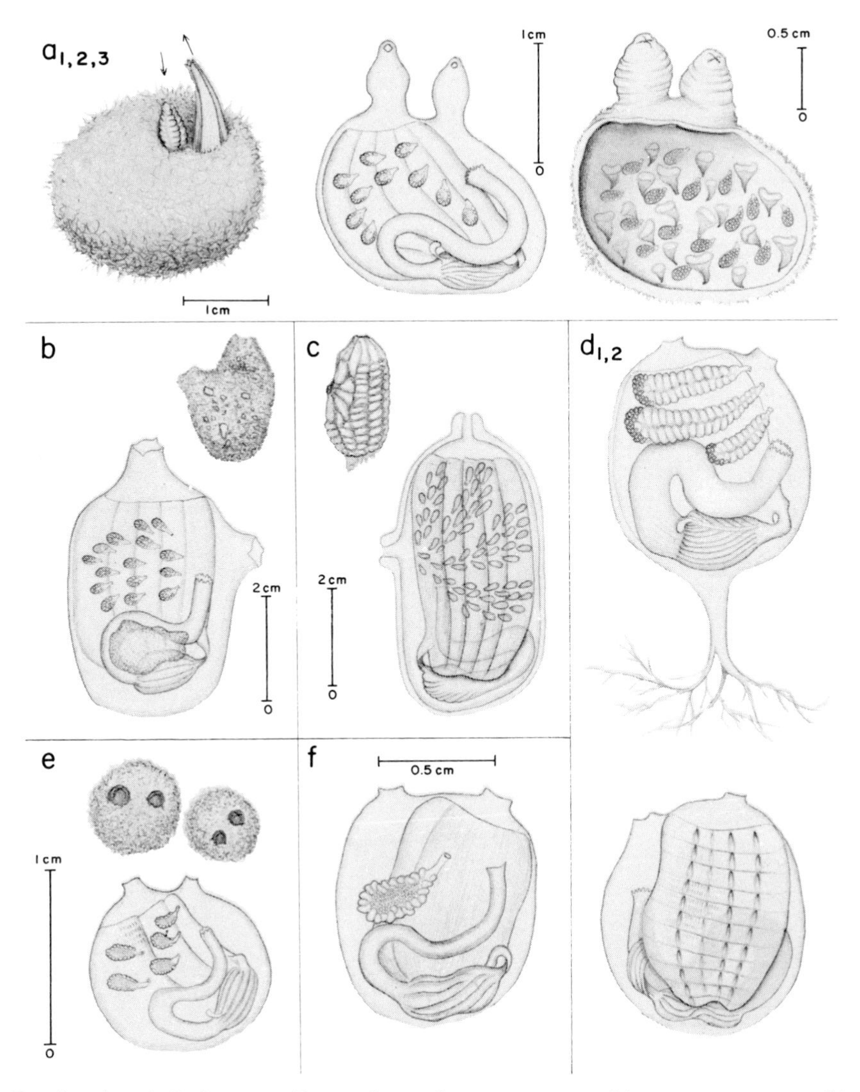

Fig. 42a–f. (a_{1-3}) *Polycarpa fibrosa* from deep water, outside appearance, zooid, and mantle inside; (b) *P. obtecta*; (c) *P. circumarata*; (d_{1-2}) *Cnemidocarpa mollis*, right and left; (e) *Polycarpa fibrosa* from shallow water; (f) *Cnemidocarpa mortenseni.*

and the test is completely attached on the ventral surface. The test is often colored red all over, or only on the siphons. In this subgenus only one true fold remains in the branchial sac. Only one tubular gonad is present on the right side. Eggs are released into the atrial cavity, and they often develop into massive tadpoles before shedding. In spite of their extremely large size, these tadpoles develop only one sense organ, the statolith. Although the similarity of these flattened, bright-colored zooids found off Cape Ann or on stones in Block Island Sound to free-living *Polycarpa* is not always obvious, their development clearly shows their evolutionary relationships to the *Styelidae.*

After examination of the typical solitary *Polycarpa fibrosa,* the rather primitive hold-over of an ancestral type in *Pelonaia,* and the flattened experiment in *Dendrodoa,* it is useful to look at the colonial *Styelidae.* As in several other families, these colonies show more simplified structure than the solitary species. They appear to have been introduced to the Atlantic shelf both from the colder northern waters and from the south into the West Indian seas. The distribution and the structure of the colonial Styelids suggests that colonial species have been derived from solitary ones several times, and in different parts of the world. The complicated budding sequences are unlike in different subfamilies and also suggest quite independent evolutionary histories.

The *Botryllinae* are found in harbors and shallow waters the whole length of the Atlantic continental shelf, especially in northern and southern portions. The bright

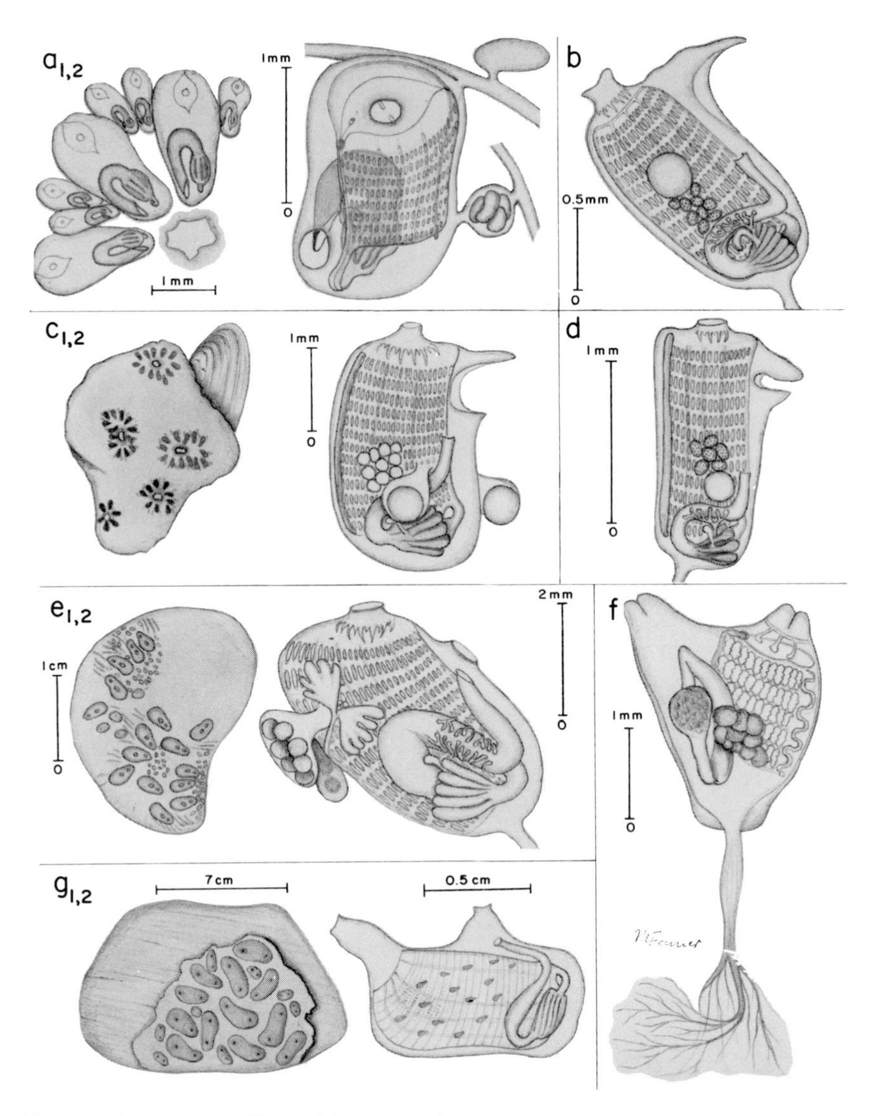

Fig. 43a–g. (a_{1-2}) *Botryllus schlosseri,* colony and zooid (after Van Name); (b) *B. planus*; (c_{1-2}) *Botrylloides aureum,* colony and zooid; (d) *B. nigrum*; (e_{1-2}) *Symplegma viride,* as found and zooid; (f) *Dicarpa simplex*; (g_{1-2}) *Polyandrocarpa maxima,* as found and zooid.

Polyzoinae are warm water species of the Florida coastal waters and the West Indies.

Botryllinae. It seems best to follow Berrill (1950) and Kott (1970) and call this group a subfamily of equal rank with the *Polyzoinae* rather than as a separate family as did Van Name (1945). *Botryllus* and one or two related genera are found as harbor-living inshore, purple-brown colonies on piles or kelp from the Gulf of Maine to Florida on the continental shelf of North America, as well as from Norway to the Mediterranean in Europe.

The best-known species is the world-wide *Botryllus schlosseri* (Plate XIV & Fig. $43a_{1-2}$). With star-like clusters of small zooids, bright purple centers and brown margins, or alternatively purple-red throughout, they attract immediate attention in all seasons. Sometimes the colors are bright red with orange margins and they change in any area from year to year. Quiescent colonies can often be found even in winter, but they grow most rapidly from June to November. They are unlike *Symplegma*, which has both oral and atrial siphons, although both form a gelatinous enclosing test. In *Botryllus* the oral siphons open separately, but the atrial siphons open into a cloacal cavity in the tunic which is shared by a number of zooids in the associated system.

In the Botryllids two buds are formed from the lower lateral portion of one zooid simultaneously and the atrial siphons remain associated. As budding continues, this process results in the formation of large systems with one shared cloacal opening in the middle. This interrelationship in a colony shows very clearly when a larva is allowed to settle and attach on a microscopic slide. If the slide is kept in a box in the sea water off a wharf, all stages in the formation of the growth of colonies can be followed and stages preserved.

Such a colony is drawn in Plate XIV, along with the finer structure in several zooids and buds. The zooids measure about 2 mm across and lie in a partly elevated position, with the oral siphon extending upward. In *Botryllus* there are no folds in the branchial sac, which has three longitudinal vessels on each side. The intestinal loop is on the left side of the branchial sac and partly below. The stomach bears a series of longitudinal lines, one of which forms a tubular cecum. The intestine opens into the atrial cavity which is a common space shared with the other zooids in this system.

The male and female gonads are found on each side of the sac. The eggs are released into the atrial or cloacal cavity where they develop to the tadpole stage. Tadpoles of *Botryllus* have only one sense organ, an otolith, but they are active and swim for a short period. Grave (1935) showed that the otolith forms some light-sensitive cells, so it is called a "photolith."

Several other growth forms of Botryllids have been described as distinct species. Van Name considered the most distinctive to be *Botryllus planus*. This forms thin encrusting sheets with flatter lying zooids. The gonads are reduced in size. This species has not been widely accepted (Fig. 43b).

Another genus, *Botrylloides aureum*, is a northern form (Fig. $43c_{1-2}$). It is found off both sides of Greenland and has migrated to deeper locations in the Gulf of Maine as far as Casco Bay. It is a darker colored species than *Botryllus schlosseri*. Its distinctive structure is a sac-like brood pouch formed by the body wall on each side. Eggs pass from the oviduct into this pouch and there development occurs. The pouch apparently aids development in northern latitudes. A darker species of the Florida Gulf coast, Bermuda and the West Indies is *Botrylloides nigrum* (cf. Fig. 43d).

In general the *Botryllinae* are a widely distributed and successful group of colonial Ascidians in the shallower and warmer waters all over the northern hemisphere. Its wide occurrence in harbors in different parts of the world suggests transport on floating objects and ships' bottoms.

Polyzoinae. This is a subfamily of colonial species, in some of which the zooids are separate except for connecting stolons. In others they appear as a surface mat of bright-colored independent zooids held together within a gelatinous test. Both oral and atrial siphon openings are separate. Gonads are attached to the inside of the mantle. Buds are formed one at a time on the lower side of the zooid under the mantle (Plate XV).

Polyandrocarpa (*Eusynstyela*) *tincta* forms bright pink or red clusters of small zooids (6 × 3 mm) attached to stones or larger Ascidians near low water mark in shore off the Florida Keys. On the Florida west coast it is often dredged from close to shore out to a depth of 20 meters. The flattened sac-like zooids are crowded in clusters, but when one is plucked out it looks very similar to one of those of the solitary species. It is clear that some of the colonial species have been derived from solitary ones quite recently.

The branchial sac has four folds on each side and there is a circle of fine tentacles with an oval dorsal tubercle. The stomach is elongate with many longitudinal folds. The gonads are attached in a double line along the mantle. There are several less common species off the west coast of Florida. One forms bright scarlet clusters: *Polyandrocarpa* (*Eusynstyela*) *floridana*.

One of these, the largest, is *Polyandrocarpa maxima*. It is found in shallow water in Apalachee Bay and southward along the west coast of Florida to the Tortugas Islands. It forms irregular masses of tough, reddish-purple, test material enclosing many completely independent zooids. These as they lie in the colony do not show any interconnecting stolons, although they have been formed in this way by buds. The bright red-brown zooids, each about 10 mm long, lie in the secreted test material with their siphons extruded at the surface. When these zooids are removed from the test, they show a structure very similar to *Polycarpa fibrosa* (Fig. $43g_{1-2}$).

Symplegma viride is another colonial genus which appears as an interconnected series of zooids overgrowing large solitary Ascidians. It appears in bright colors: gray, light green, purple, or orange. The zooids are embedded in a transparent gelatinous tunic, rather like *Botryllus*, and they are interconnected by vascular tubules. But each of the two siphons of *Symplegma* opens separately on the surface and there is no grouping of the zooids into symmetrical systems within the test. The branchial sac is without folds in this species, but there are four longitudinal vessels with about six straight stigmata in each row between them (cf. Plate XV & Fig. $43e_{1-2}$).

The stomach has about twelve fine, longitudinal ridges with one large pyloric cecum. The intestine shows many fine intestinal tubules attached in one area opposite the stomach. The gonads consist of two testis lobules and one large ovary on each side. *Symplegma* is found in deeper water off the west coast of Florida at 15 meters.

Huus (1937) suggested the very ancient origin of *Symplegma* by showing that it is found off the northern shores of Australia, off the East Indies, along the African east coast, and in the Red Sea, then in the waters of

Bermuda and the Gulf of Mexico, but in no waters between. Perhaps it was spread around in mid-Mesozoic times several hundred million years ago when the North Sea and the Indian Ocean were closer together and interconnected and so missed dispersal in the South Atlantic.

Styelinae. The third large subfamily of *Styelidae* is the *Styelinae*, and it includes many different species of solitary Ascidians. These show adaptations in many different directions. Like the colonial species just reviewed, all the solitary species could have been derived from the generalized structure shown by the primitive deep water form *Polycarpa fibrosa* (see p. 86). The evolutionary diversity of the *Styelidae* in different genera appears greater than that shown by either of the other two families of the *Stolidobranchia*, the *Pyuridae*, and the *Molgulidae*.

On the southern portion of the Atlantic continental shelf off the coast of Georgia and Florida the northern cold water, mud water favoring species, *P. fibrosa*, is replaced by *Polycarpa obtecta.* This larger yellow-brown species lives only partly buried in sand bottoms of shallower water. The zooids may measure 40 mm and may be attached loosely by the ventral side. A rather similar subspecies, *Polycarpa spongiabilis*, was described by Traustedt, but it cannot be separated from the previous species on the North American shelf (Plate XVI & Figs. 18, $42a_{1-2}$ & e).

A larger, brown-red or scarlet-patched species, *Polycarpa circumarata*, is found in shallow water off the northwest Gulf coast of Florida, especially in Apalachee Bay. It may be 60 mm in length and scarlet through its tufted length. These two large species show how a deep, mud-living northern and more primitive *Polycarpa fibrosa* has given rise to larger and more sluggish species on the southern Atlantic continental shelf.

In addition, this genus has become established in deeper water of the continental slope of North America at depths from 1500 to 5000 meters. *Polycarpa albatrossi* was found by Verrill and called a *Molgula.* It was restudied by Van Name (1912) and has been again described by C. and F. Monniot (1970), along with a number of new species from specimens of the Woods Hole Oceanographic Institution's survey of the deep water toward Berumda.

A number of *P. albatrossi* specimens have been dredged by the R. V. *Gosnold* collecting for the U.S. Northern Marine Fisheries Laboratory and the Woods Hole Oceanographic Institution. These specimens were found at deep water locations as follows: (1) south of Martha's Vineyard; (2) south of Long Island; (3) in Hudson Canyon; (4) off Cape May; (5) off Delaware; (6) off Virginia. The depths from which these specimens were taken were all about 2500 meters. This is very deep water, but it seems probable that all specimens came from the continental slope.

These specimens need further, more careful study, but they resemble those described by Van Name in external appearance, rather than those described by the Monniots from the Bermuda Deep collection. It appears that the species shows greater variability along the North American continental shelf than the Monniots describe. It appears that the original Van Name description fits our specimens and is more correctly called *Polycarpa albatrossi*, than is the Bermuda variety described by the Monniots.

The genus *Cnemidocarpa* is only slightly more specialized than *Polycarpa*, and its species have become adapted to similar environmental niches on the Atlantic continental shelf of North America. *Cnemidocarpa* appears to have migrated southward from the boreal zone, especially along the North American continental shelf. It is of interest that one of the largest and most successful of such species, *Cnemidocarpa verrucosa*, is exclusively antarctic in distribution. Presumably, it was transferred to the seas off the antarctic continent in that early Mesozoic time when that was associated with the land masses of India and Africa.

Cnemidocarpa prefers sand rather than mud and is thinly covered by fine sand grains. A few hair-like filaments growing out of the test on the ventral side hold the zooids in place on the bottom. It has already been pointed out that *Cnemidocarpa mollis* often attaches its filaments to other small Ascidians, especially *Cratostigma singulare* and *Bostrichobranchus pilularis*, which are living in the same situation but are not so securely anchored against the marine current. *Cnemidocarpa* differs from *Polycarpa* structurally in possessing fewer but longer tubular gonads with ovary and testes enclosed in one membrane.

Cnemidocarpa mollis appears as small, sand-covered, elongated sacs 20 mm long with two short square siphons close together on the dorsal side. The branchial sac has four folds and about twenty straight stigmata in rows between the folds. The stomach is broad with about eighteen fine longitudinal folds. The intestinal loop is short. There are four vial-shaped gonads in the wall on the left, and six or more on the right. The white ovaries are internal with a large oviduct and surrounded by brown testis lobes with ducts to the siphon (Figs. 19 & $42d_{1-2}$).

Cnemidocarpa mollis has migrated southward along the North American continental shelf from its probable boreal source. It is found in smooth areas on the sand a half mile to five miles off shore in the Gulf of Maine to Cape Cod. There are also additional occupied areas with sandy bottoms on Georges Bank. Recent dredging shows it is present on the sand at depths of 10 to

30 meters where there is a moderate current as far south as Sapelo Island, Georgia (see Fig. 5). The fact that earlier dredging records recorded this species only as far south as Cape May, suggests that it has moved southward just as *Molgula arenata* has done. The distribution which our collecting has shown indicates that *Molgula arenata* may be found also in deeper water than *Cnemidocarpa mollis*.

Just as was described for *Polycarpa*, there is a small deep water-related *Cnemidocarpa* species called *Cnemidocarpa mortenseni* (Figs. 19 & 42f). This has been found on the continental shelf off Georges Bank at depths of 80 to 100 meters and possibly in deeper water east of Long Island. This smaller species differs from *C. mollis* in possessing a tough brownish tunic and an elongate stomach. The branchial sac shows only one fold, and there are three large gonads with a central ovary surrounded by lobed testis.

Just as with *Polycarpa*, the distribution of *Cnemidocarpa* indicates that these small Styelids which feed continuously on tiny particles carried in the off-shore coastal currents, starting in the boreal, have spread south along the Atlantic continental shelf of North America.

The genus *Styela* is so well known by those who scrape the surface of piles or stones just below tide levels along the northeast coast, that it gives its name to the family. It includes several species found sparsely in shallow waters in many parts of the world and some related genera found in deeper spots on the continental slopes. Two of the species of *Styela* found just beyond tide water are the most successful and structurally most advanced members of the family on the Atlantic continental shelf. Both are a little more susceptible to shore pollution than are the *Molgula* species with which they compete, so that in such spots only *Molgula* survives. In general, however, when sea squirts are mentioned on the Atlantic coast of the United States, two species only are likely to be produced. They are *Styela partita*, north of Cape Hatteras, and *Styela plicata* to the south.

Styela partita is the irregular, brown, tuberous species about 25 mm long with an extended and striped aperture. It is found in shallow water from Cape Ann to the West Indies. It is especially common north and south of Cape Cod, in quiet harbors, but it is not found north of the continental shelf of the United States. *Styela plicata* is much larger, 60 to 80 mm or more, but it is limited to warmer waters off North Carolina, and southward. It is common off many of the West Indies. It is smoother, lighter, often cream-colored, compared with *S. partita*, with broad conspicuous furrows and rounded tubercles, and lobed, striped apertures, and it grows at least three times as large (Figs. 16, 17, & 44a & b).

In spite of wide distribution close to shore of *Styela partita* all along the continental shelf of the United States, it is not found north of Nova Scotia. Yet a very similar, if not identical, species is found close in along the coasts of Europe and the Mediterranean. It seems of interest that although it is not found off North America, one other Styelid species has a very extensive distribution on a world scale. This is *Styela nordenskjoldi* as described by Kott (1969). This species ranges all around the antarctic continent, but in much deeper water. It seems clear that the roots of this antarctic *Styela* are different from those on the American shelf and that the genus *Styela* has had a complex evolutionary history.

It was suggested earlier that *Polycarpa fibrosa* of moderately deep boreal waters was the most probable ancestor of the *Polycarpa* species of farther south and shallower water closer to shore. It is of interest that much the same species distributions hold for the genus *Styela*. There are at least two species which are circumpolar in cooler, moderately deep waters, and which range south of Nova Scotia. One is the rather small elongate tubercle-studded *Styela coriacea* (Fig. 44d). When dissected it shows two curved gonads, one on each side with the testis tuft prominent. The branchial sac has four folds with large stigmata. The intestinal loop is large and crowded together. Several specimens have been dredged south of Cape Ann. The other species, which is found in the St. Lawrence estuary and occasionally in the Gulf of Maine, is *Styela rustica*. This is a furrowed elongate species 30 to 40 mm long with a point or spine between the dorsally placed siphons. Either of these two species, especially *S. coriacea*, might have been the source in the remote past of the shallow-water Styelids of the North American and European shelves.

The two major species of the Atlantic continental shelf in-shore distribution should have a brief structural description. *Styela partita* has a rough, rather leathery tunic. It is attached at its ventral end and often by one side in addition. The shape is often irregular due to crowding. The branchial sac has four prominent folds, with large straight stigmata. The stomach is long, with prominent ridges. There are two gonads on each side composed of a long ovary and many prominent testis lobes attached by fine ducts. Eggs are large and clear. They form small tadpoles after being shed into the sea water to develop. Some years ago (1921) Van Name described a very similar but larger *Styela atlantica*. It occurs only in deep water on the continental slope in the Atlantic, but also in the Pacific Ocean (Fig. $44c_{1-2}$).

The southern coastal species, *Styela plicata* is very similar to *S. partita*. It is often massive in size, with specimens from Apalachee Bay measuring over 100

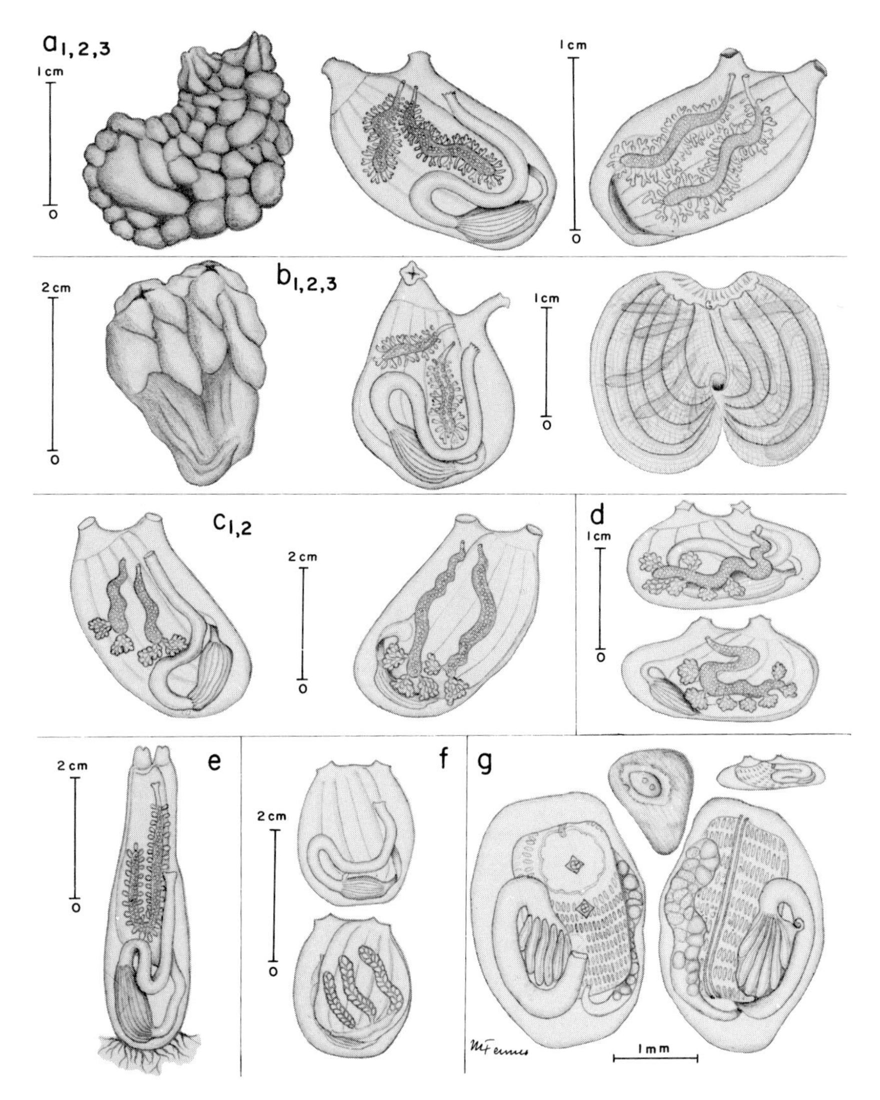

Fig. 44a–g. (a_{1-3}) *Styela partita,* outside, right and left; (b_{1-3}) *S. plicata,* outside, right, inside; (c_{1-2}) *S. atlantica,* right and left; (d) *S. coriacea,* right and left; (e) *Pelonaia corrugata*; (f) *Dendrodoa pulchella*; (g) *D. carnea.* (After Van Name.)

mm. Often in the same shallow sites may be found the southern coastal *Molgula occidentalis. Styela plicata* often shows a white furrowed test, which makes it the most easily recognizable solitary Ascidian in the region.

Recently, a species new to the Atlantic continental shelf has been found in Narragansett Bay by H. W. Pratt, a staff member of the National Marine Fisheries Service. It is *Styela clava,* a narrow-bodied zooid averaging 60 mm in length, with the lower test drawn out in a stalk attached to the bottom. When removed from the test, there are two reddish gonads on the left side, and five or six on the right, similar to *Styela plicata.* This species is found in the Pacific Ocean, especially off Japan and Korea, and it also has been found by Millar in the Atlantic, south of Britain. The discovery of *Styela clava* in Narragansett Bay records another species found in moderately deeper water in both the Atlantic and the Pacific.

One final abyssal species shows a fairly successful deep water adaptation in the Styelid family. Three specimens were collected by R. V. *Gosnold* (8/64) for the Woods Hole Oceanographic Institution east of Hudson Canyon and in Baltimore Canyon. Another was collected east of Cape Hatteras by R. V. *Eastward* for Gilbert Rowe (1968), then of the Duke Marine Laboratory. The depths were 1408 meters to 2335 meters for *Gosnold* collections, and 1955 m for the *Eatward* specimen. The specimens are all *Dicarpa simplex.* The species was named by Millar from collections of the Danish *Galathea* expedition. It is clear that this is a world-wide deep sea species, since it is found also in the South Atlantic and the Indian Ocean.

Dicarpa simplex is a triangular zooid about 4 to 5 mm wide, borne on a pedicle twice that length. Although the species lives in the complete darkness except for phosphorescent species, it showed a pink and greenish color in zooid and stalk. In the mounted specimens it is shown that the branchial sac has no stigmata and appears like a net. Its wide meshes allow the flow of sea water carrying oxygen and food particles. Although it is at reduced levels, the sea water in deep water still carries oxygen in solution. From the sea water at the surface above these deep water sites, there is a continuous descent of organic particles. Oxygen is transferred from the sea water into the blood vessels in the mesh of the branchial sac, and food particles are collected in the mucus roll in the cavity of the branchial sac (Figs. 21 & 43f).

The structure shows that the safety of quiet deep water involves provision of other necessities. The gonads are massive. Eggs are released from the atrial siphon and apparently develop close to the parental zooid in the deep water to parental size without forming tadpoles.

The species count shows eighty-eight species of *Ascidiacea* living on the Atlantic continental shelf of the United States, At least two of those mentioned are undescribed, and a more careful account will be published later. Doubtless there are some others, but they will be species holding on in live habitat groups, in small numbers. Some general suggestions concerning probable changes in species distribution in the long past and the immediate future will be given in the final chapters.

VII. Distribution in Relation to Continental Drift

An occasional resurvey of the geographic distribution of all the species making up any large group of animals or plants is desirable in order to determine whether geographic stability is being maintained. It makes it possible to find out if one or more species are supplanting others, or if the environment is undergoing changes which stimulate increases in the areas covered by one or more previously studied biologic groups. For example, our recent study showed that the small *Molgula arenata* and the rather large bright red *Polycarpa circumarata* are both found in much broader areas off the southern states than has been recorded previously. Going back much further in time, it was in 1911 that Hartmeyer initiated a series of such studies on Ascidian species found off Europe, and these studies were extended through 1923. Van Name always listed the geographic distribution of all the species in his successive studies of American Ascidian species, but without the addition of new geographic surveys. A most active and informative reexamination of the distribution of European Ascidian species is being made by the independent studies of Francoise Monniot and Claude Monniot as in 1965 and in several later publications.

The chief advantage of a more accurate examination of the geographic distribution of all the species in the small class *Ascidiacea* on the North American Atlantic shelf lies in its evolutionary implications. It makes it possible to suggest how the various species reached their present locations and the possible changes which might take place in the years ahead as a result of human activities originating on shore. The *Ascidiacea* are an ancient group of marine benthic animals widely spread over the continental shelves of all continents. They are attached to the bottom, where they remain throughout their adult life. Their food consists of particles brought by the sea water currents. In addition to the environmental conditions at the sites where they live, the success of a species may be influenced by colony formation. Colonies may be formed by budding from the initial egg-developed zooid, so that a mass of individuals all derived from the same original egg remain associated. Species movements are largely restricted to those determined by egg release in the relatively slow ocean currents by which they are surrounded. In addition, some spreading movements occur as a result of the release of swimming tadpoles. Slow as these species movements are, they result in changes and transitions of common species even in the course of fifty- to hundred-year periods commonly separating species distribution studies. But it seems probable that environmental movements are the most effective agents in species distribution changes.

All of the species seen and identified in this most recent study of *Ascidiacea* of the North American Atlantic shelf are listed in Table III. For each species an average depth at which zooids were found is noted. This is one of the most important environmental determinants of the success of any species, whether solitary or colonial. Shallow water depth locations were found for fifty-one species, medium depths were commonest in thirty-five species, and only two among those identified were deep water species. From the table it can be seen that in the order *Enterogona* about three-fourths of the species live in shallow water, but in the more advanced *Pleurogona* the number of species having a deeper water preference is closer to one-half. Most of the species, like *Dicarpa simplex*, which have become adapted to deep water living of over 300 meters are members of the more advanced families of the *Mogulidae* and *Styelidae*.

But the most useful comparisons derived from a study of Ascidian species distributions are not from depth records but the geographic spread of North American species, that is, how many are found in North America only, and how many are found also on continental shelves of Europe and Asia. Such comparisons make it possible to guess at the movements of Ascidian species in the last eighty million years of the earth's history since the continental split which in Mesozoic times gave rise to the Atlantic Ocean. To aid in this consideration, a more complete tabulation of species distributions had been made of thirty-nine of the commonest Ascidians of the Atlantic continental shelf. These are more extensive distribution comparisons than have been made

Table IV. Distribution of 39 Common Species of *Ascidiacea* on the Continental Shelves of Atlantic North America and of Europe

Family & subfamily (1)	Genus & species (2)	Location on American shelf (3)	Location on European or other shelf (4)	Gathering of similar locations into groups (5)
Order *Enterogona*				
Suborder *Aplousobranchia*				
Cionidae	*Ciona intestinalis*	Boreal & No. Temperate	Northern & temperate north. hemisphere	(I)
Clavelinidae *Clavelininae*	*Clavelina oblonga*	Carolinas to Florida	So. Europe & Mediterranean	(VII)
Polycitorinae	*Eudistoma olivaceum*	West Florida	(Not found)	(VI)
	Cystodytes dellechiajei	Gulf of Mexico Off Florida	Warm seas off So. Europe & Asia	(Pair) (VIII)
Holozoinae	*Distaplia clavata*	Northeast—shallow	(Not found)	(II)
	Distaplia stylifera	Georgia, Florida, Gulf of Mexico	Southern off Europe & Asia	(VIII)
Polyclinidae	*Aplidium pallidum*	Gulf of Maine—deeper water	Off No. Europe—deeper water	(III)
	Aplidium constellatum	Northeast & southern	(Not found)	(V)
Didemnidae	*Didemnum albidum*	Gulf of Maine—middle depths	Off No. Europe & Asia circumboreal	(I)
	Didemnum candidum	Gulf of Maine to Gulf of Mexico	Shallow water off Europe	(VIII)
	Trididemnum tenerum	Northern, medium depths	Off Europe, circumboreal	(I)
	Diplosoma macdonaldi	Southern seas, shallow	Related sp. off Europe, medium depths	(III)
Suborder *Phlebobranchia*				
Perophoridae	*Perophora viridis*	Shallow water, temperate to warm	Related sp. in warmer water off Europe & Asia	(VIII)
	Ecteinascidia turbinata	Warmer water, shallow	(Not found)	(VI)
Corellidae *Rhodosomatinae*	*Rhodosoma turcicum*	Temperate to warm	Off So. Europe & Asia	(Pair) (VIII)
Corellinae	*Chelyosoma macleayanum*	Northern cool water	Off Europe, circumboreal	(I)
	Corella borealis	Northern, medium depth	North Europe	(II)
Ascidiidae	*Ascidia prunum*	Northern cool medium depth	North Europe	(II)
	Ascidia callosa	Cool water	North Europe	(II)
	Ascidia nigra	Southern, Georgia, Florida	Red Sea & off Southern Asia	(VIII)

(Table IV continued)

Family & subfamily (1)	Genus & species (2)	Location on American shelf (3)	Location on European or other shelf (4)	Gathering of similar locations into groups (5)
Order *Pleurogona*				
Suborder *Stolidobranchia*				
Pyuridae				
Bolteniinae	*Halocynthia pyriformis*	Northeast, Gulf Maine	Off Europe, shallow water	(II)
	Boltenia echinata	Northeast, medium depths	Circumpolar	(I)
Pyurinae	*Pyura vittata*	Southern, shallow, also West Indies	Indian Ocean, off southern islands	(VIII)
Heterostigminae	*Cratostigma singulare*	Off Cape Cod, & Cape Cod Bay	Reported east of Azores	(IV)
Molgulidae				
Molgulinae	*Molgula citrina*	Northeast, Temperate	Common, shallow water	(II)
	Molgula manhattensis	Northeast common, rare to Gulf of Mexico	Similar related species in Europe	(VII)
	Molgula occidentalis	Southern Ga. to Gulf of Mexico	Rare off So. California	(VII)
	Molgula siphonalis	Northeast, shallow to medium depth	European, similar	(II)
	Molgula arenata	Gulf of Maine to Ga. on sand	(Not found)	(V)
Eugyrinae	*Bostrichobranchus pilularis*	Gulf of Maine to Fla. west coast	(Not found)	(V)
Styelidae				
Polyzoinae	*Polyandrocarpa maxima*	Fla. West coast & Islands	East Indies Tropical Asia & off Brazil	(Pair) (VIII)
	Symplegma viride	So. Car. to Gulf of Mexico	Tropical Asia, East Indies	(Pair) (VIII)
Botryllinae	*Botryllus schlosseri*	Cape Cod to Gulf of Mexico	Temperate to warm, world wide	(III)
Styelinae	*Polycarpa fibrosa*	Northeast in cold current & medium depths	Europe similar	(II)
	Polycarpa albatrossi	Deep water, off mid-American slope & Bermuda Trench	Deep water, Atlantic Trench	(VI)
	Cnemidocarpa mollis	Northeast to Fla. shallow to medium	Europe rare, in warm water only	(III)
	Styela partita	Northeast common in shallow water	Europe rare	(III)
	Styela plicata	Southern, common Ga. to Brazil	Mediterranean, Indian Ocean, & East Indies	(VIII)
	Dendrodoa pulchella	Northeast, esp. toward north on rocks in cold current	Europe similar northern only. Circumboreal	(I)
	Total species checked	....	39	

Note: See Summary in Table V.

Table V. Summary of the Distributional Data of Table IV Suggesting that Ascidian Species Were Placed by Continental Drift from Mesozoic Time

Group numbers from Table IV	Total species distri-butions	Geographic ranges of species groups found	Summaries & origins	Approximate time scale
(I)	6	North America, Europe, & North Asia, circumboreal		
(II)	8	North America & Europe, & boreal distribution	19 species: One source spread to Europe & North America by continental drift	Cretaceous ± 80 million years
(III)	5	North America & Europe. No boreal persists		
(IV)	1	North American shelf only. Northern	7 species: One source but species persisted in North America only after continental drift	Late Cretaceous ± 65 million years
(V)	3	North America only. Northern & southern		
(VI)	3	North America only. Southern		
(VII)	3	North America, & similar species present off South Europe	13 species: One source spread by continental drift. Species died out in Europe after transport through Tethys Sea from Europe to the Indian Ocean	Eocene ± 50 million years
(VIII)	10	North America, southern only. Similar species are found off Southern Europe or off tropical Asia		

Total species checked—39

up to this time, and they are shown in Table IV and Table V.

In Table IV, the taxonomic position of each of the thirty-nine species selected is shown by order and family, and examples are given for each family. In column (3) are noted the geographic areas on the continental shelf of North America which include the species range. The preferred locations within this area, as shallow sand or deep gravel bottom are not indicated, but they were usually noted where the species was described in Chapter VI. Column (4) notes for each species, whether the same or a nearly related species is found on the European continental shelf. In addition, it is recorded in column (4) whether the species has been identified from southern European waters—the Mediterranean Sea or the Indian Ocean. The broad regions of species distribution for each of the thirty-nine species on the North American Atlantic shelf, together with its distribution on European continental shelves are in the last column (5) in Table IV. In this last column, species distributions that are broadly similar are grouped together for comparison and are numbered with Roman numerals (I–VIII). These species distribution patterns appear to justify broader conclusions, involving mechanisms and a time scale of species transport which has not been previously apparent.

To aid in interpreting the data assembled in Table IV, column (5), the group totals have been transferred to Table V and aggregated into two additional columns there, suggesting certain obvious conclusions. Table V, column (1), shows the groups numbers of Table IV repeated at the left, and the numbers of species which were found to fall into each are totaled in the second column. In the third column of Table V, the distinctive geographic distributions of each of the groups are summarized. Reviewing them briefly we find that group (I) includes North American species which are circumpolar in the northern hemisphere. This means that the species in group (I) are found off North America and also on sea-covered land masses from points off Greenland, Iceland, North Europe, and Asia. Thus their ranges include the continental shelves of North America and Europe, as well as northern or arctic Asia.

Group (II) shows species with a rather similar distribution to (I), except that North America and Europe are interconnected in ranges by specimens found in the somewhat more southerly boreal zones, such as Iceland and Davis Strait. Group (III) includes species which have been collected off North America and Europe in approximately similar temperature zones, but no specimens from intervening under-water shelf areas have been found. As will be noted in the fourth column in Table

V, it is reasonable to add the numbers in groups (I), (II), and (III). This shows that nineteen of the thirty-nine listed species are found in rather similar underwater shelf zones on both North American and European continental shelves. The most reasonable explanation of the presence of nineteen species on the shelves of both continents is that they were distributed at the same time, and there has been little or no species movement. Only the continents have been separating through late Mesozoic time.

Continuing the examination of Table V, it will be seen in addition that group (IV) includes only one species. This is the unusual Pyurid, the tiny *Cratostigma singulare*, currently found off the angle of Cape Cod, and in southern Cape Cod Bay. This species is said to be present southeast of the Azores Islands, but I could not find it in 1973, with unsatisfactory dredging equipment. Next are groups (V) and (VI). All of the species in these three middle groups are found on the North American shelf only. If we add these three groups together the total number of species is seven, about one-sixth of the total number listed in Table IV. The most plausible explanation of this kind of species distribution is like the previous one, but the species persisted in North America and died out in Europe.

Finally, there are two groups remaining. Group (VII) contains three species found in the southern portion of the shelf of the eastern United States, and these same species can be dredged from warm sea bottoms in southern Europe. Group (VIII) contains ten species which are now found on the continental shelf of the southern United States and also off southern Europe, but also on Asiatic undersea land masses, as in the Red Sea or even still farther south into the East Indies. One of the most widely cited examples of these pairs of the same species on the Atlantic continental shelf, and in shallow water in the Red Sea and the Indian Ocean, is the common species *Symplegma viride*. It is found off both coasts of Florida and off the Texas coast as far as Padre Island. It is said to be present on the margins of the Indian Ocean as far as Java. These extraordinarily widely separated pairs need to be more carefully studied. The last two in Table V include thirteen species, or a final one-third of the total number of Ascidian species picked out for distribution study. Together the five columns tabulated in Table V suggest that the thirty-nine selected species found on the Atlantic continental shelf have been subject to several different influences in time and place before being brought to their present geographic locations.

If it is reasonable to assume that the class *Ascidiacea* was in existence on some continental shelves in early Paleozoic times, there have been many surface movements on the globe to which living marine species were exposed. One of the most recent of these surface influences is the onset of the Ice Age—a relatively recent change in the northeastern portion of the Atlantic continental shelf brought about by the melting and recession of the last of the ice caps about 10,000 years ago. In North America the ice extended south over most of the Gulf of Maine and Cape Cod and Long Island. As it receded so much ice was melted that the sea level was elevated 100 to 300 feet. This flooding greatly extended the Atlantic continental shelf, not only in the Gulf of Maine but farther south, except where there was compensating land movement (cf. Milkman & Emery 1968). The correctness of the flooding hypothesis has been amply demonstrated by the finding of peat deposits on the bottom at Georges Bank in the Gulf of Maine (Emery, Wigley, & Rubin 1965) and the dredging up of teeth and bones of arctic mammoths from the shelf farther south.

Other much more ancient geological events have probably influenced the distribution of Ascidian species along the Atlantic continental shelf. Shifts in the polar axes have occurred in the course of Paleozoic time. These might have been induced by shifts in the position of the magnetic pole, as cited by Bain (1966). More influential events occurred in Eocene times when the Tethys Sea separated Europe and Asia and so joined the Indian Ocean to the North Atlantic. By this throughway the colonial Styelid *Symplegma viride* and others of the so-called "pairs" listed in Table IV, column (5), may have been transferred from the Indian Ocean to the developing Atlantic. So they reached even the West coast of Florida and the Gulf of Mexico. These species apparently persisted at both ends of their original ranges and were eliminated by other earth movements in between.

It was Huus (1937) who first commented on these curious distributional pairs cited in Table IV. We are drawn inevitably into a consideration of the relation between the still earlier continental changes and the present distribution of many more Ascidian species, as already suggested above. These implications of additional relationships between continental movements and Ascidian distributions become a little more obvious with consideration of the facts summarized in Table V.

Table V column (4) shows that Ascidian species distributions on the Atlantic continental shelf fall into three major classes. The first is the largest and contains nineteen of the thirty-nine species listed, or one-half of the total. All the species in this group are also found somewhere on the continental shelves of Europe. Some of these species have been dredged also from the shelves of intervening boreal land masses such as Iceland or Greenland.

The second class listed in Table V column (4) contains seven species, all of which are found only on the North American shelf. These may extend in distribution from north to south along the United States, or they may be restricted to either. This class represents one-sixth of the total.

The third class in Table V column (4) contains thirteen species and represents one-third of the total. It includes only southern species usually from shallow waters, and they are found also off southern Europe. As already discussed, it is this group which includes the so-called "identical pair" species which are found along the North American shelf and also off southern Asiatic shelves, but with no interconnecting coastal distributions between.

In the past it has been customary to explain the distribution of many Ascidian species on the continental shelves of both Europe and North America by assuming that some species were transported from European to American sites by migration in the recent past. Those found in Asiatic locations were supposed to have moved initially to Europe and from there to have migrated to North America by way of the underwater shelves of Greenland and other intervening land masses. None of these suggestions take into account the more recent views on continental movement in the northern hemisphere, which must have had a major influence on the distribution of Ascidian species as well as the location of other benthic animals. It now seems very probable that the presence of many of the same species on European and North American continental shelves may be due not to the movements of the animals but rather as a result of changes in the positions of the major continents on the world globe as a result of what is now usually called "continental drift." It is widely held that in late Paleozoic time all the present-day continents were more or less fused into a single massive land mass aggregated in the far Southern hemisphere of the globe (Gondwanaland).

By mid-Mesozoic time this great continental aggregation had partially separated into a northern and a southern portion, with several breaks forming partially enclosed ocean bodies. One of these in a later time period interconnected southern and northern oceans through the European continent and was known as the "Tethys Sea." By the Cretaceous Age the Atlantic Ocean was progressively forming and enlarging. As it extended northward it split South America from Africa and, eventually, North America from Europe. The northward widening of the Atlantic from east to west continued and resulted in some progressive shrinking in the Pacific.

The broad overall changes in the continental forms and their movements are simply and helpfully delineated in the figures and the brief outline of Dietz and Holden in *The Journal of Geophysical Research* (September 1970). This is a condensed summary of current views and adapts their figures, originally appearing in "This Brittle Planet" in *The Sciences* (November 1970). Their figures are reproduced in Figure 45, by permission of the editor and author.

It seems quite certain that there were Ascidian species living a benthic existence on the continental shelves of the world's continents long before the massive continental drifting of the Mesozoic. It is probable that Ascidian fossil specimens are known from an early time period, as *Ainiktozoon* from upper Silurian of Britain (Scourfield 1937) and certainly from late Pliocene (F. Monniot 1970).

By late Mesozoic times both shallow water Ascidian species, and some of medium water depths, were living along the margins of the continental masses, as suggested in the figure. As the splitting divided the land masses, water-living animal species were carried along on the separating continental shelves of both continental masses. The Ascidian species which had originally lived on the southern continental shelves became inhabitants of the continental shelves of both of the separating continents. Thus these slow continental movements furnish a plausible basic explanation for the great similarities between the Ascidian species found off the coasts of Europe and North America, as well as often on the boreal zones between. Most of the Ascidian species common to both continents did not migrate from the European continental shelves across the boreal zone to the shelf of North America. Actually, these benthic species, for the most part, remained where they were and were carried by the land on which they lived. Only very slowly they may have extended the shelf areas on which they were attached as the surrounding environments changed.

In order to make the conclusions suggested it will be helpful at this point to examine a number of distribution patterns of common Ascidian species made from National Marine Fisheries Laboratory records, with some additions from the years since 1968. The species for which the records are shown are: *Aplidium glabrum, Ascidia obliqua, Boltenia ovifera, Molgula siphonalis, Molgula citrina, Molgula manhattensis, Polycarpa fibrosa, Styela partita.* An outline sketch of each of these species is given in Figures 28c, 32c, 33a, 39a, 36d, 36a, 42a, and 44a. The distribution of some of these and other species is given in Figures 5 and 10.

The facts shown by the distribution records can be summarized very briefly. For the *Enterogona* species, *Aplidium glabrum* has been dredged out of the Gulf of Maine from south of Nova Scotia in the Bay of Fundy, close to shore off Eastport, and south to waters off Cape Ann. It occurs also in deeper water north of

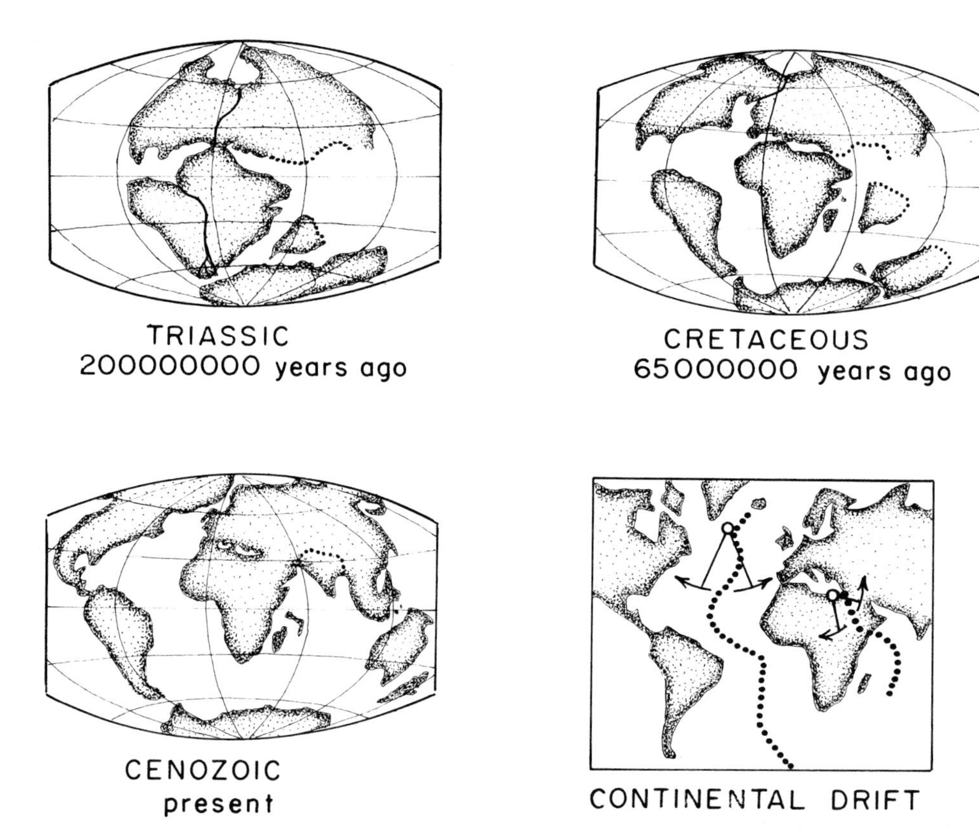

Fig. 45. Continental Drift. Relation of the continents during the Triassic, the Cretaceous, and at present. At the end are given the lines along which expansion of the oceans is continuing, and the apparent hinge points of this movement.

Georges Bank and on Nantucket Shoals. The locations in Europe, as reported by Hartmeyer (1923) and Berrill (1950), are very similar. Northwest France and southwest England distributions are difficult to account for as steps in migration from Europe to North America, but rather easier if they are thought of as being passively carried from a common source in continental drift.

The next species to be checked is the larger solitary *Ascidia obliqua* with a rather roughened test and a smaller alimentary loop than other deep water *Ascidia* species. It is only in water over 100 meters deep that it is regularly found, and it gets no closer to shore than Georges Bank. In northern European waters it is widely distributed and is dredged south to the latitude of southern Scotland, and it is also found in deep water. This species can be easily identified by its distinctly grouped stigmata and curved papilla. Its similar distribution in European and North American waters suggests that it started at a common source and the source divided.

Passing on to the families of *Pleurogona* in Table IV, the first species for which distribution in North American waters and those of northern Europe can be compared is *Boltenia ovifera*, the stalked species, and the other species often found with it, *Boltenia echinata*. These species are found circumpolar in arctic or subarctic regions and can be found in southerly latitudes, especially on the North American shelf. The stalk of the former and the spines of the latter make them both easily identifiable, and they are often attached in association. Their very similar distributions on the two continental areas, and even in northern Pacific waters, strongly suggests that each of the regions was colonized at the same time and only later did the areas drift apart.

The North American distribution of two additional *Pleurogona* species, each found off European and North American coasts, can be cited in order to make it clear that the similarities in distribution noted for boreal species may appear also in species found farther south of the northern boreal zone. *Molgula citrina* is found in shallow American coastal areas around Cape Cod and south in Long Island Sound. It is common in European waters in similar water regions but farther north. Similar but much wider occupation regions and closer to shore occur for *Molgula manhattensis* in America and Europe, assuming that the European *Molgula ampul-*

loides is the same. For these species the numbers in shore areas off Massachusetts, Rhode Island, and in Long Island Sound are very much larger than those found off European coasts. Nevertheless, the zooids are as similar as many neighboring North American specimens. In each the egg develops into a tiny tadpole fully formed and functional, but without an ocellus. Such close resemblance certainly suggests each was derived from a common ancestor on the shore of a splitting continent.

A similar explanation best fits the American and European off-shore distribution of both *Cnemidocarpa mollis* and *Molgula arenata*. Each of these species is found loosely attached on fine, hard sand where there is an active water current. Both can be dredged from moderately deep water north of Cape Cod, but *Cnemidocarpa* usually prefers deeper water of 20 meters. *Molgula arenata* is one of the most numerous Ascidian species ten miles off shore on the Georgia continental shelf at about 10 meters depth. Although also present off northern European coasts, each of these species of sand-dwelling Ascidians is present on both sides of the Atlantic in northern latitudes. One could make similar comparisons between the American and the European distributions of two additional species, *Polycarpa fibrosa* and *Styela partita*. Each of these species may very well have been present in the fork of the continental split during Mesozoic times.

One final case of unusual or unexpected distribution of a rarer species may be cited before leaving the question of possible interrelationship between present distribution and continental movement. *Chelyosoma macleayanum* specimens have been found several times from off Cape Ann rather close to many tightly attached specimens of *Dendrodoa carnea*. *Chelyosoma* has been collected also on the Pacific side of North America above the Bering Strait, so it is usually said to show circumpolar distribution. It is strange, nevertheless, to find that the *Albatross* dredged several deep water specimens of what must be a closely related species from deep water in the latitude of Panama. Perhaps these species, with the spiral stigmata like their related *Corella*, may have been located between the Indian and Atlantic oceans when the Indian was interconnected with the growing Atlantic.

The probable effects of the movements of the continents on Ascidian species distribution seem much more relevant than was the disappearance of the Pleistocene ice cap in the northern continents during the last fifteen thousand years. During that period the ice receded from Long Island, Cape Cod, and eastern Canada. The most obvious change was the elevation of sea level by about 100 meters. So great a deepening resulted in the covering of most of the continental shelf in the northeast and also farther south. This resulted in much under water shelf available on which Ascidians and other benthic animals could become attached, so perhaps there was an increase in total Ascidian abundance closer to shore. It is quite obvious at the present time that off Eastport, Mt. Desert Island, and Cape Ann many Ascidian species are found which in the region about Cape Cod or farther south can be dredged only in deeper water.

With the description of each species of *Ascidiacea* found in this new survey on the Atlantic continental shelf some statement of its distribution is given. Usually this has agreed with previous published descriptions, but in several cases it has been found that the distribution has been extended, usually from north to farther south. More than half of the Ascidian species are found in the Gulf of Maine and, for example, *Aplidium pallidum* and *Polycarpa fibrosa* still remain there. But *Cnemidocarpa mollis* and *Molgula arenata* are now found as far south as the shelf off Georgia, and possibly still farther.

In addition to the northern species there are a number of species found from Georgia southward to Florida and around the peninsula along the west coast of Florida. Many of these are regularly distributed in the West Indies region. Good examples of species native to southern continental shelf areas are *Styela plicata* and *Molgula occidentalis*. There is possible basis for believing that some southern species, like *Distaplia bermudensis*, have migrated away from the continental region. But some of the Ascidian species found on the continental shelf off Florida east and in the Gulf of Mexico came there from the West Indies and farther south. This may have been the migration pathway for species like *Ecteinascidia turbinata* and *Eudistoma olivaceum*.

In addition to species of the northern and southern regions of the Atlantic continental shelf, there are a number of species of the northern portions of the Atlantic shelf which are found in similar locations on the continental shelf of North America and Europe. Some exceptional species are circumpolar, as for example *Molgula citrina* and *Boltenia echinata*. More of them are collected off both continents without any known intervening locations, like *Ascidia prunum* and *Molgula retortiformis*. It has been already suggested that such split species distributions have their origin in the progressive separation of Europe and North America during the Mesozoic era, with the formation of the Atlantic Ocean (see Table IV and Table V).

The distributional data cited in this chapter leads to the rather unexpected conclusion that many species of *Ascidiacea* are very old. The separation of North America from Europe as a result of continental drift must have occurred late in Mesozoic time, perhaps Cretaceous, close to seventy million years ago. This indicates that such species as *Distaplia clavata* or *Boltenia echinata* or *Molgula complanata* have changed so little in seventy million years that the same species name can be used. This would not be true for land-living insects. Nevertheless, insects preserved as fossils in amber are recognizably close to modern species. Also, F. Monniot's fossil *Cystodytes* is the same genus as the living species, and *Symplegma viride* lives off Java and West Florida, and the two groups have been separated fifty million years.

The suggestion seems reasonable that Ascidian species are more constant than the coasts on which they live. Apparently they live on the continental shelves without the action of species-changing evolutionary natural selection such as acts more rapidly on fishes, or land-living reptiles and mammals.

VIII. The Ascidian Larva and the Beginning of the Chordate Line

Among the different major branches or phyla of animals there are four which are frequently grouped together because of a common pattern in development. This does not mean that they show resemblances in present adult structures, but rather that in development they show a common pattern in the formation of the coelomic cavity by out-pocketing of the primitive gut in the gastrula stage. There are two such pouches which form the cellular lining of the body cavity, and their tissues become the mesoderm, the source of muscles and skeleton of the adult (cf. Fig. 46).

This method of formation of the coelom is found in the development of at least three major groups of animals: the *Echinodermata* (sea stars and sea urchins); the *Chaetognatha* or *Enteropneusta* (acorn worms); and the *Chordata*, including *Tunicata* (sea squirts and salps), *Cephalochordata* (Amphioxus), and, finally, the *Vertebrata*, which includes jawless *Agnatha* and jaw-carrying *Gnathostomata* of water and land.

The phylum *Echinodermata* is made up of a strange assortment of widely different spiny marine animals including sea stars, sea urchins, sea cucumbers, and sea lilies which are very successful. The *Chordata* are also a numerous and successful phylum among the nektonic life in the sea. It seems difficult to assume that similarities in the formation of the coelom can be used to suggest ancient relationships between the phyla. Nevertheless, the likeness in developmental processes is seen not only in enterocoel pouches but in the schemes of formation followed in the embryology of digestive glands, circulatory systems, and nervous tissue development. All indicate distant relationships. These resemblances have justified grouping the *Echinodermata*, *Chaetognatha*, and *Chordata* into one assemblage which is given the name *Deuterostomea*. The term means little more than that all these animal groups must have had a common stem ancestral stock in the long past.

This broad view of ancestral relationships can be justified by the observation that there are several distinctive patterns in larvae in the different phyla and classes of Deuterostomes which resemble each other, although the adults are a long way apart. In Figure 46 are sketched two common larvae in different orders of *Echinodermata*, the Gastrula and Pluteus of sea urchins, the Bipennaria of sea stars, and the Tornaria of acorn worms. Similar larvae such as those sketched here suggest a common relationship. This means that in the distant past sea stars and acorn worms were descended from an ancient common stem form, not that one adult was derived from the other. Even so good a zoologist as W. Garstang (1928) defended the unjustified view that Ascidians were descended from Tornaria-like larvae. Others have defended the view that the Tornaria of acorn worms were derived from Echinoderms. It is now realized that the constant recurrence of similar larval patterns in the development of Echinoderms, Chaetognaths, and Ascidians means that in the long past all of these shallow water living groups were derived from a common Deuterostome stem.

With a rather similar developmental pattern each produces larval forms broadly alike but eventually changing, so that each grows to a different group of adults. The Echinoderm gastrula forms a Pluteus or Bipennaria larva, depending on its genetic determinants. The Tornaria develops into an acorn worm. The Ascidian egg shows the Chordate pattern in early development (Fig. 47) and then hatches into a distinctive larva, the Ascidian tadpole. This shows the distinctive arrangement of a dorsal notochord and a tiny nerve cord above in every one of the Ascidian species. It is so uniform in what seems to be the most ancient group that it is reasonable to characterize the *Ascidiacea* as the probable ancestors of the Chordate phylum, which includes the Vertebrate class as its newest addition since the Cambrian.

The *Ascidiacea* are generally considered the most probable ancestors of the *Chordata* because their eggs develop into the unique tadpole larvae. This is an elongate zooid with a head attached to a tail-like posterior portion. The head contains the nervous system rudiment, with a vesicle carrying two ganglia, an ocellus or visual nerve center, and an otolith which is an equilibrium or auditory center. In addition, it includes the undeveloped beginnings of an intestine and of a circulatory

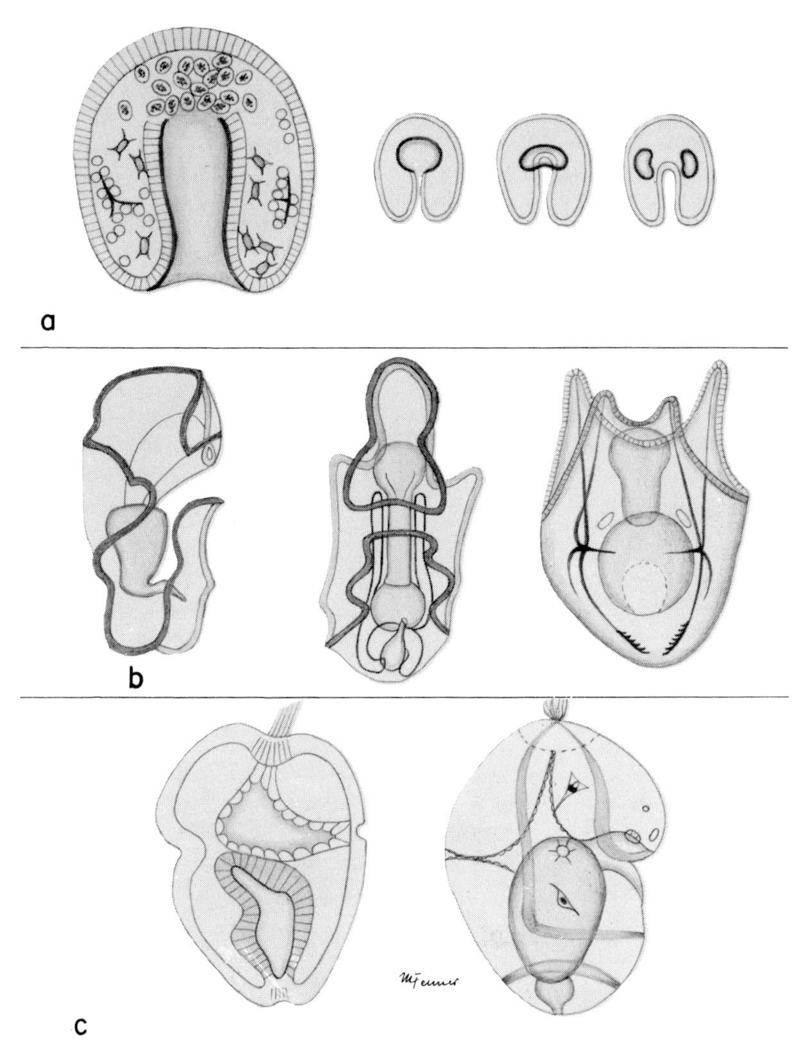

Fig. 46a–c. Early Development. (a) Echinoderm gastrula and formation of enterocoel pouches. (b) Pluteus and Bipennaria. (c) Tornaria larvae of Enteropneust.

system. Behind the head stretches the tail, which is stiffened by a long, rod-like notochord with a thin nerve cord dorsal to it. Rudimentary muscle tissues enable the larva to vibrate back and forth to swim soon after it goes into the sea water. It was the discovery by Kowalewsky in 1867 of this tadpole larva, suggestive of an ancient frog tadpole, which gave the first evidence that the Ascidians might be the ancient forerunners of the vertebrates (cf. Fig. 48, *Ascidia mentula*).

The Ascidian tadpole has undergone certain evolutionary changes in the course of the Paleozoic time since it first appeared, perhaps in the early Cambrian time. It is possible that the earliest larva was symmetrical with a wide tail fin like that of *Ciona* (Fig. 49a). Later the tail became more fin-like and the organs in the head developed further as in *Halocynthia* (Fig. 49b). Another generalized type of tadpole is that shown in longitudinal section in Figure 50 which is *Molgula citrina* redrawn from Caswell Grave. This is a larva about one millimeter long which may swim for several hours before attachment at the head end. The head has one black sense organ, the statolith, and the developing but not yet functional mouth and branchial sac. The semi-rigid notochord passes from the head through the whole length of the tail, and the very thin nerve cord lies dorsal to that. There are three tiny lateral muscle bands along either side of the notochord, which stimulate the side-to-side beating of the tail under nervous control. The larval mouth is not broken through and the larva cannot feed. The section should be compared with the similar swimming larva of *Molgula complanata* (Fig. 37b), which was from a specimen collected inside Padre Island, Texas, by Nancy Rabalais.

A study of two additional longitudinal sections of

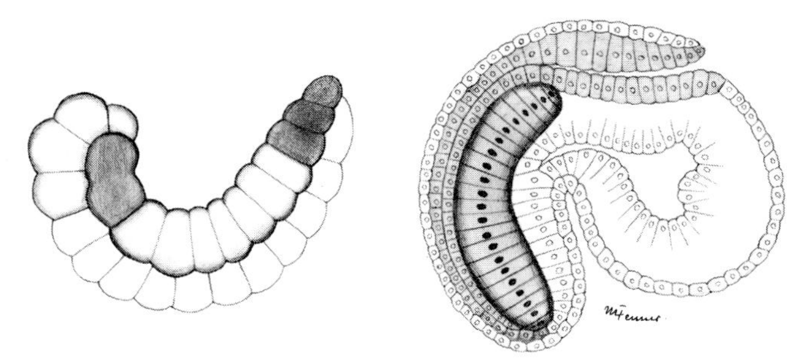

Fig. 47. Ascidian gastrula and larva.

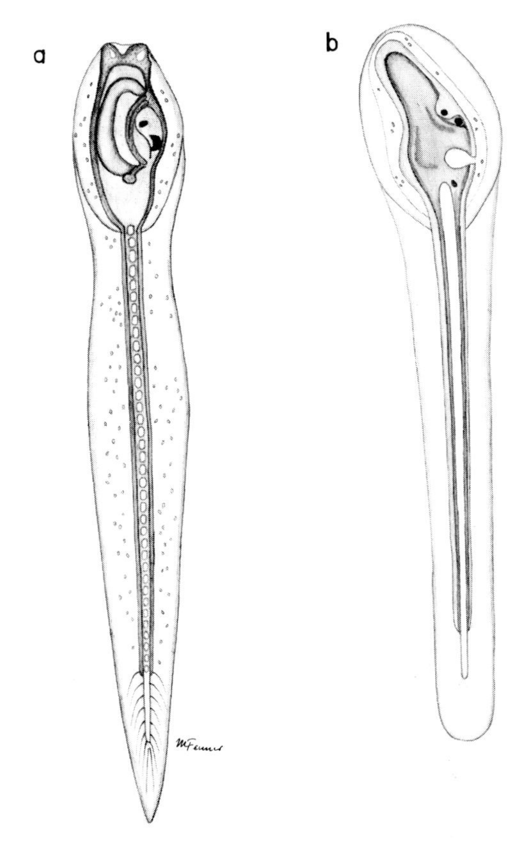

Fig. 49a–b. Ascidian tadpoles. (a) *Ciona* (after Millar); (b) *Halocynthia* (after Berrill).

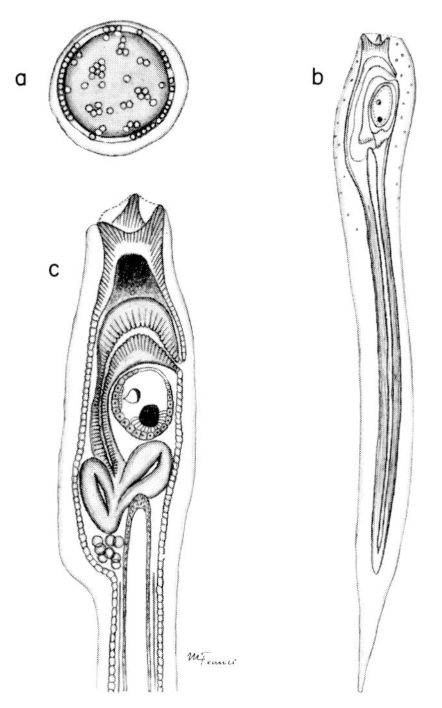

Fig. 48. *Ascidia mentula* tadpole, first described by Kowalewsky in 1867. (a) egg; (b) longitudinal section; (c) detail of head.

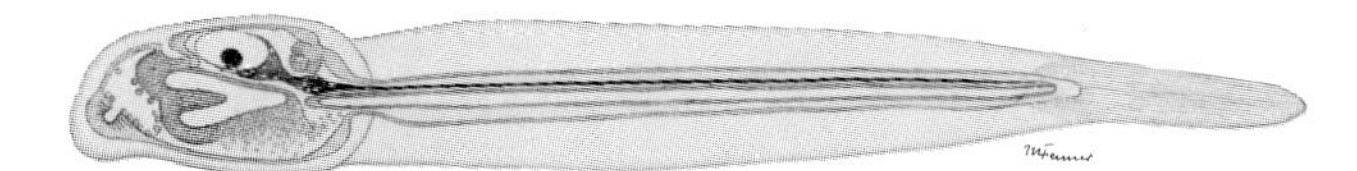

Fig. 50. *Molgula citrina* tadpole in longitudinal section (after Caswell Grave).

Ascidian tadpoles, *Aplidium constellatum* (Fig. 51a) and *Perophora viridis* (Fig. 51b as drawn by Grave), shows the typical basic pattern, as in the *Ciona* larva. In both the typical chordate arrangement appears, but the tail with its dorsally placed notochord and nerve cord is longitudinally flexed at right angles to the plane of the head. Each larva shows the normal Chordate arrangement just as it appears in the unflexed larva of *Ascidia, Ciona, Halocynthia,* or *Molgula.* All of these Chordate larvae resemble each other in that they swim briefly, although they never become feeding zooids, and then settle down and undergo metamorphosis to a sac-like mollusc-like adult with complete loss of their Chordate plan. None of these Ascidian tadpoles ever becomes a sexually mature Chordate adult following a paedogenetic transition. There have been several studies in which Ascidian larvae were subjected to changed environmental treatment, as by Berrill (1929), in order to induce maturity in the tadpole stage, but they have always eventually led to metamorphosis. Nevertheless, it seems probable that there were early Ascidian species which formed tadpoles in early Palaeozoic time before there were any living vertebrate precursors.

If it is impossible to find acceptable evidence that an Ascidian tadpole has ever given rise to a sexually mature Chordate adult, is there any other evolutionary pathway possible which was open during early Paleozoic times? It seems now that there may have been

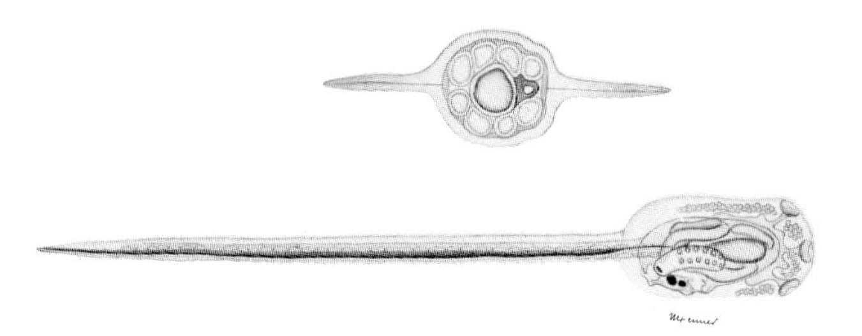

Fig. 51a. *Aplidium constellatum* tadpole, including cross section (after Caswell Grave).

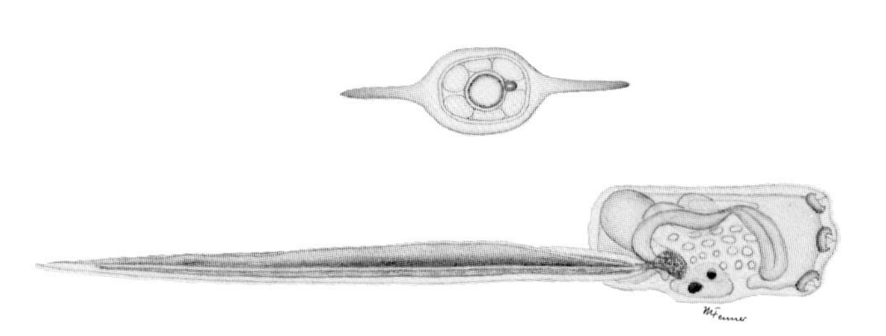

Fig. 51b. *Perophora viridis* tadpole, including cross section (after Caswell Grave).

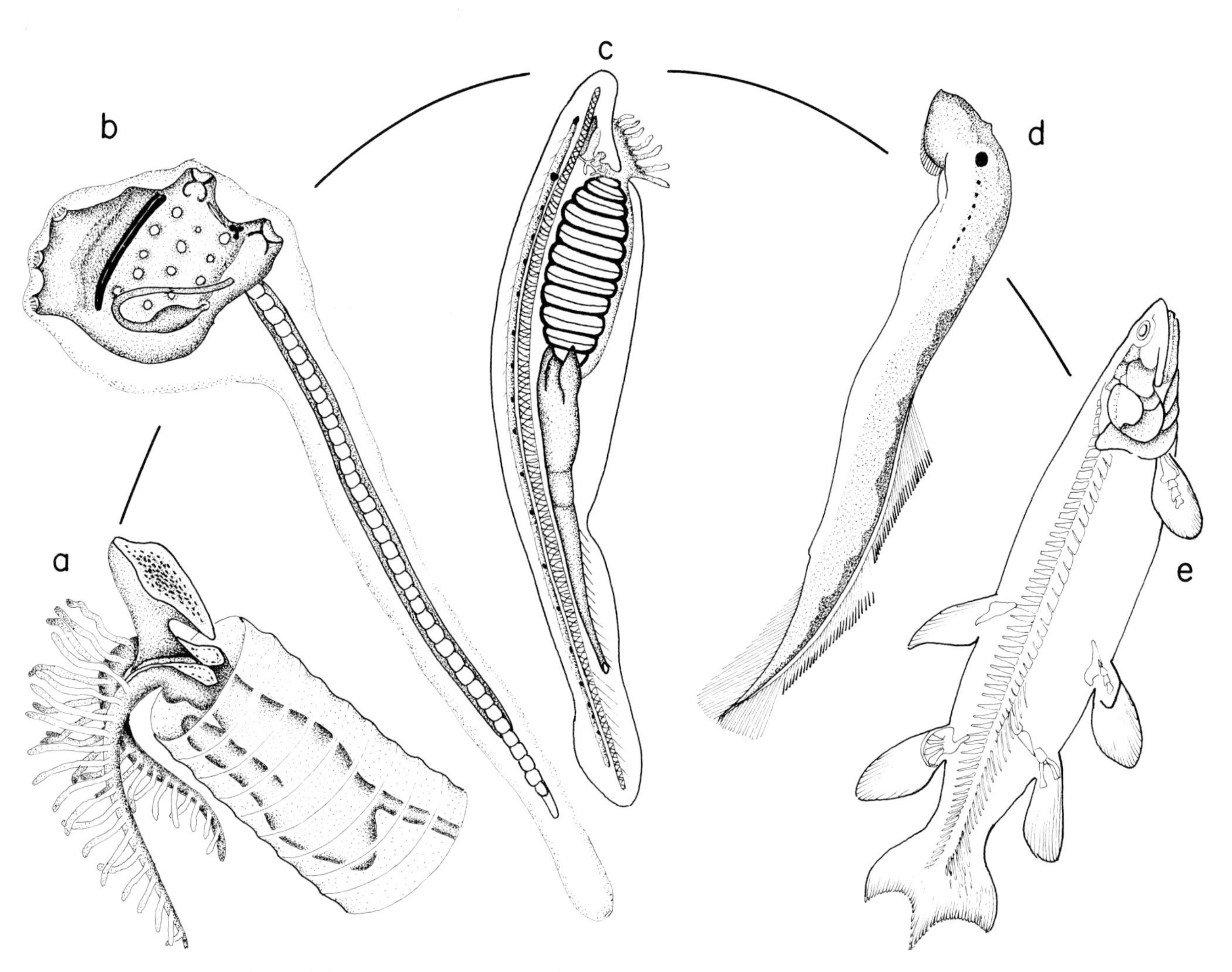

Fig. 52. Living examples of possible steps in the evolution of the vertebrates. (a) *Rhabdopleura*, a pterobranch hemichordate; (b) Ascidian tadpole; (c) *Amphioxus*; (d) Lamprey eel (jawless); (e) Fish (jaws, backbone). (After Romer)

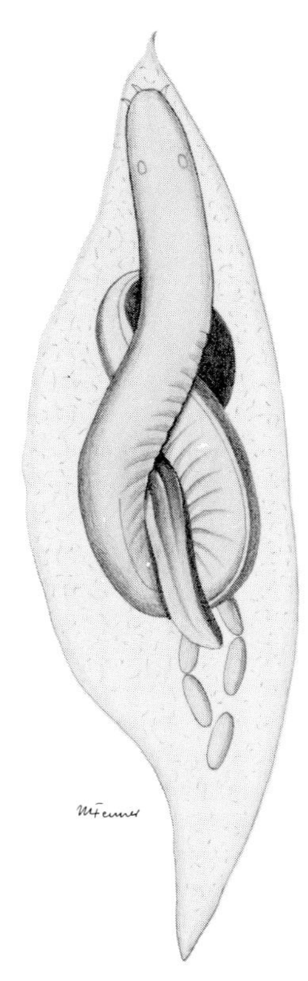

Fig. 53. *Bdellostoma stouti* (after Bashford Dean).

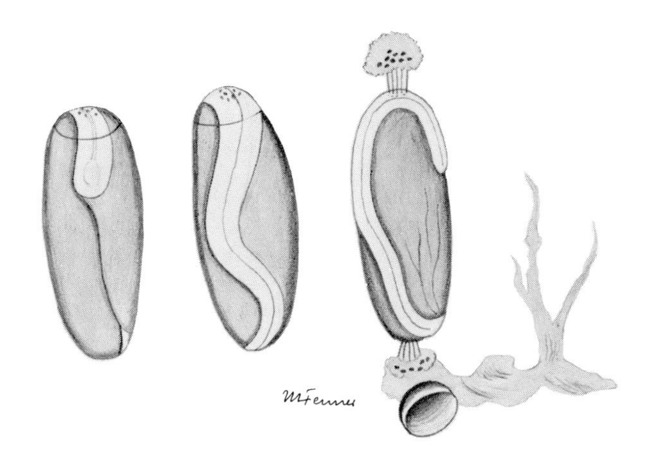

Fig. 54. Developing eggs of *Bdellostoma stouti* (after Bashford Dean).

developing early in the Paleozoic another line of ancestral Chordates which may have been more promising as a stem line leading to the precursors of the early fishes. Instead of forming small eggs shed into the sea water and developing into tadpoles at an evolutionary dead end, it now appears that there may have been a primitive fish-like species which formed large eggs retained and developed within the parent's body. It has long been assumed that fish-like ancestors would have been too late in time to initiate such an evolutionary transition. With the continuing studies of a number of vertebrate paleontologists it has become possible, and indeed probable, that in early Paleozoic seas there were small fish-like species resembling some of present day *Cyclostomata*. The living species form rather large eggs which carry on development within the parent's body. When the developing tadpole or fish-like zooid is large enough to feed it is released from the parent's body. The yolk of the original large egg remains attached to the under side of the developing embryo until long after hatching. The little fish-like larvae have open mouths through which they gradually feed while they continue to use up the attached yolk. They are Chordate larvae with attached food supply.

Perhaps the living species which best illustrates the transition is the marine hagfish, *Bdellostoma stouti*. The development of *Bdellostoma* was described seventy-five years ago by Bashford Dean (1899). The living hagfish is a scavenger with a cartilage skeleton. Several of Dean's figures have been redrawn in Figures 53, 54, and 55. The large eggs are shed into the sea water and carry the developing embryos on their surface. When hatched the young *Bdellostoma* carries the remaining yolk attached to its under side. It swims for some days while feeding and at the same time gradually absorbing the remainder of the egg yolk. In this stage it resembles the Ascidian larva and is a functional living tadpole.

There are a good many fish-like Vertebrate fossils with a heavy armor of bony plates from early Paleozoic, perhaps even Ordovician. They were small fish-like species called Ostracoderms, and they may have been ancestral to the Vertebrates, but the most likely ancestors were the ancient Cyclostomes themselves. Studies of several small fossil species of *Cyclostomata* by Stensio (1968) show that some of these species were living very early in the Paleozoic, certainly in Ordovician, and perhaps even in the Cambrian. Such great age makes these species contemporaneous with the early Ascidians. Indeed, it suggests that there were some of these *Cyclostomes* present as vertebrate forebears at the same time that early Ascidians were forming the various tadpole larvae and metamorphosing to the sac-like adults. The Chordates were the last of the major phyla of animals to be initiated, but it seems necessary to conclude that in the earliest epoch of the Paleozoic their structural plan was established, and there were fish precursors developing on large eggs.

It is of general interest to examine the beautifully illustrated book, *Ascidians of Sagami Bay* (Japan) by H. Hattori and T. Tokioka, editors (Iwanami Schoten,

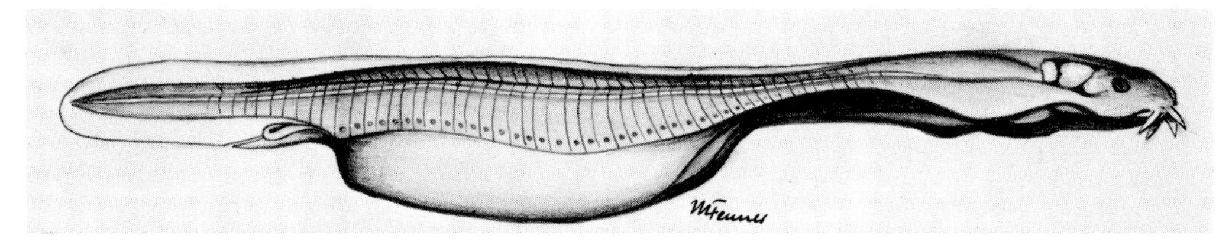

Fig. 55. Embryo of *Bdellostoma stouti,* with yolk sac (after Bashford Dean).

Tokyo, 1953). In this book are found many pictures and descriptions of Ascidian specimens collected by His Majesty, the Emperor of Japan. There are numerous black and white drawings, and three full-color plates. In Chapter I a series of diagrams is shown in which the reverse view concerning the evolution of Ascidians from that set forth in the previous paragraph is defended, and in Chapter II, it is suggested that early jawless vertebrates might have degenerated to produce after a long time Ascidian-like descendants. This reversed evolutionary view seems hardly plausible to most zoologists, but it not necessary to accept it. Rather marine zoologists can appreciate the industry and enthusiasm, as well as the long critical study, which have gone into the preparation of this distinguished Japanese volume. Similar studies in many parts of the oceanic environment all over the world are needed to make complete the story of the distribution of *Ascidiacea.* Our summary of the Ascidians of the Atlantic continental shelf of North America adds another chapter to the catalogue.

Appendix. List of Species of *Ascidiacea* Found on the Atlantic Continental Shelf of the United States, 1968-1974. (Total Number of Species–88)

Family & subfamily	Genus & species	Place found
Order *Enterogona*		
Suborder *Aplousobranchia*		
Cionidae	*Ciona* Fleming 1882	
	C. intestinalis (Linnaeus) 1767	Northern harbors
Clavelinidae		
Clavelininae	*Clavelina* Savigny 1816	
	C. oblonga Herdman 1880	Southern shelf
	C. picta (Verrill) 1900	
	C. gigantea Van Name 1921	
Polycitorinae	*Eudistoma* Caullery 1909	Southern shallow
	E. capsulatum (Van Name) 1902	
	E. olivaceum (Van Name) 1902	
	E. hepaticum (Van Name) 1921	
	E. tarponense Van Name 1945	Gulf of Mexico
	E. carolinense Van Name 1945	
	Cystodytes von Drasche 1884	
	C. dellechiajei Delle Valle 1871	
Holozoinae	*Distaplia* Delle Valle 1881	
	D. clavata (Sars) 1851	Boston Bay
	D. bermudensis Van Name 1906	Southern
	D. stylifera (Kowalewsky) 1887	
Polyclinidae		
Polyclininae	*Aplidium* Savigny 1816 (*Amaroucium*) Milne Edwards 1841	
	A. pallidum Verrill 1871	Gulf of Maine
	A. glabrum Verrill 1871	
	A. pellucidum Leidy 1855	
	A. stellatum Verrill 1871	
	A. constellatum Verrill 1871	North & south
	A. bermudae Van Name 1902	Southern
	A. exile Van Name 1902	
Didemnidae	*Didemnum* Savigny 1816	
	D. albidum Verrill 1871	Northern
	D. candidum Savigny 1816	North & south
	D. vanderhorsti Van Name 1924	Southern
	D. amethysteum Van Name 1902	
	Trididemnum Delle Valle 1881	
	T. tenerum (Verrill) 1874	Northern
	T. savigny (Herdman) 1886	North & south
	T. orbiculatum (Van Name) 1902	Southern

Family & subfamily	Genus & species	Place found
	Diplosoma Macdonald 1859	
	D. macdonaldi Herdman 1886	Southern
	Lissoclinum Verrill 1871	
	L. aureum Verrill 1871	Northern
	L. fragile (Van Name) 1902	
	Echinoclinum Van Name 1902	
	E. verrilli Van Name 1902	Southern
Suborder *Phlebobranchia*		
Perophoridae	*Perophora* Giard 1872	
	P. viridis Verrill 1871	
	P. bermudensis Berrill 1932	
	Ecteinascidia Herdman 1880	
	E. turbinata Herdman 1880	Florida
	E. conklini Berrill 1932	Florida Keys
	E. tortugensis Plough and Jones 1939	Tortugas, Fla.
Corellidae		
Rhodosomatinae	*Rhodosoma* Seeliger 1855	
	R. wigleii sp. nov.	Cape Ann, Mass.
Corellinae	*Chelyosoma* Broderip & Sowerby 1830	
	C. macleayanum Broderip and Sowerby 1830	Gulf of Maine off Cape Ann
	Corella Alder & Hancock 1870	
	C. borealis Traustedt 1886	Northern Gulf of Maine
Ascidiidae	*Ascidia* Linnaeus 1767	
	A. prunum Mueller 1776	Gulf of Maine
	A. callosa Stimpson 1852	Medium depth
	A. obliqua Alder 1863	
	A. interrupta Heller 1878	Georgia shelf
	A. nigra (Savigny) 1816	South
	A. curvata (Traustedt) 1882	
	A. sp. uncertain	Off Cape Ann
Order *Pleurogona*		
Suborder *Stolidobranchia*		
Pyuridae		
Pyurinae	*Pyura* Molina 1782	
	P. vittata (Stimpson) 1852	Georgia south
	Microcosmus Heller 1878	
	M. exasperatus Heller 1878	Tortugas Is., Fla.
Bolteninae	*Boltenia* Savigny 1816	
	B. ovifera (Linnaeus) 1767	Gulf of Maine
	B. echinata (Linnaeus) 1767	
	Halocynthia Verrill 1877	
	H. pyriformis (Rathke) 1806	
Heterostigminae	*Cratostigma* F. Monniot 1961	
	C. singulare (Van Name) 1912	Off Cape Cod
Molgulidae		
Molgulinae	*Molgula* Forbes & Hanley 1848	
	M. citrina Alder & Hancock 1848	Vineyard Sound
	M. complanata Alder & Hancock 1848	Shallow
	M. siphonalis Sars 1859	Gulf of Maine
	M. griffithsii (Macleay) 1825	Gulf of Maine
	M. (*Meristocarpus*) *retortiformis* Pizon 1899	Nantucket Sound
	M. manhattensis (DeKay) 1843	Vineyard Sound
	M. provisionalis Van Name 1945	Buzzards Bay
	M. robusta (Van Name) 1945	
	M. occidentalis Traustedt 1883	Georgia & south
	M. arenata Stimpson 1852	Cape Cod to Florida
	M. lutulenta (Van Name) 1912	Deep water

Family & subfamily	Genus & species	Place found
Eugyrinae	*Bostrichobranchus* Traustedt 1883	
	B. pilularis (Verrill) 1871	North and south
	Eugyra Alder & Hancock 1870	
	E. arenosa padrensis new subspecies	Southern Texas
Styelidae		
Botryllinae	*Botryllus* Gaertner 1774	
	B. schlosseri (Pallas) 1766	North and south
	B. planus (Van Name) 1902	Southern
	Botrylloides Milne Edwards 1841	
	B. aureum Sars 1851	
	B. nigrum Herdman 1886	Southern
Polyzoinae	*Symplegma* Herdman 1886	
	S. viride Herdman 1886	Georgia south
	Polyandrocarpa Michaelson 1904	Florida, west
	P. maxima (Sluiter) 1904	
	P. (*Eusynstyela*) *tincta* (Van Name) 1902	
	P. (*Eusynstyela*) *floridana* (Van Name) 1921	
Styelinae	*Polycarpa* Heller 1887	
	P. fibrosa (Stimpson) 1852	Gulf of Maine
	P. obtecta Traustedt 1883	Georgia, south
	P. albatrossi (Van Name) 1912	Bermuda deep
	P. circumarata (Sluiter) 1904	Shallow, West Fla.
	Pelonaia Goodsir & Forbes 1841	
	P. corrugata Goodsir & Forbes 1841	North
	Dicarpa Millar 1955	
	D. simplex Millar 1955	Shelf slope
	Cnemidocarpa Huntsman 1913	
	C. mollis (Stimpson) 1852	North to south
	C. mortenseni (Hartmeyer) 1912	
	Styela Fleming 1882	
	S. partita (Stimpson) 1852	North to south shallow
	S. plicata (Leseuer) 1825	Southern
	S. atlantica (Van Name) 1912	
	S. coriacea Alder & Hancock 1848	Northern
	S. clava Herdman 1882	Narragansett Bay
	Dendrodoa McLeay 1824	
	D. pulchella (Verrill) 1871	Northern Gulf of Maine
	D. (*Styelopsis*) *carnea* (Agassiz) 1850	Gulf of Me. to Block Island

References

Allee, W. C. 1923. Studies in marine ecology. Pt. 1. *Biol. Bull.* 44:167–91. Pt. 3. *Biol. Bull.* 44:205–53.

Ärnbäck-Christie-Linde, A. 1922–34. Northern and arctic invertebrates in the collection of the Swedish State Museum. I–XIII:Tunicata. *Kungl. Svenska Vet. Akad. Handl.* (3):1–9.

———. 1924. A remarkable pyurid tunicate from Novaya Zemlya. *Archiv fur Zool.*, no. 15, 16:1–7.

Bain, G. W. 1966. A Laurasian looks at Gondwanaland. Annex to vol. 67. *Geol. Soc. of South Africa*, pp. 1–37.

Barrington, E. J. W. 1965. *The Biology of Hemichordata & Protochordata.* University Reviews in Biology. San Francisco: W. H. Freeman and Company.

Beneden, P. J. van. 1847. Recherches sur l'embryogénie, l'anatomie, et la physiologie des Ascidies simples. *Nov. Mem. Acad. Sci. Belgique* 20:1–66.

Berrill, N. J. 1928. The identification and validity of certain species of Ascidians. *Jour. Marine Biol. Assoc.* 15: 159–75.

———. 1929. Studies in Tunicate development. I. General physiology of development of simple Ascidians. *Phil. Trans. Royal Soc.* London B, 213:37.

———. 1931. Studies in Tunicate development. II. Abbreviation of development in the Molgulidae. *Phil. Trans. Royal Soc.* London B, 219:281–346.

———. 1936. Studies in Tunicate development. V. Evolution and classification. *Phil. Trans. Royal Soc.* London B, 226:43–70.

———. 1948. The nature of the Ascidian tadpole with reference to *Boltenia echinata. Jour. Morph.* 82:269–85.

———. 1950. *The Tunicata, with an Account of the British Species.* London: The Ray Society, pp. 1–354 (excellent bibliography).

———. 1955. *The Origin of the Vertebrates.* London: Oxford University Press.

Brien, P. 1948. *Tunicata.* In Grasse, P., *Traité de Zoologie* 11:1–553.

Child, C. M. 1922. Developmental modification and elimination of the larval stage in the Ascidian *Corella willmeriana. Jour. Morph.* 44:467–514.

Conklin, E. G. 1905. The organization and cell lineage of the Ascidian egg. *Jour. Acad. Phila.* 12:1–119.

Dean, Bashford. 1899. *On the Embryology of* BDELLOSTOMA STOUTII. Festschrift 70te Geburtstag v. Carl von Kupffer, Jena: Fischer.

Dietz, R. S., and John C. Holden. 1970. This brittle planet. *The Sciences*, New York Academy Sciences, no. 11, vol. 10: 8–11.

Emery, K. O., R. L. Wigley, and Meyer Rubin. 1965. A submerged peat deposit off the Atlantic Coast of the United States. *Limnology and Oceanography* 10 (suppl. November):R97–R102.

Garstang, W. 1928. The morphology of the Tunicates, and its bearings on the phylogeny of the Chordata. *Quart. Jour. Micr. Sci.* 72:51–187.

Giard, A. 1872. Recherches sur les ascidies composées ou synascidiens. *Arch. Zool. Exp.* 1:501–704.

Goldberg, E. D., W. McBlair, and K. M. Taylor. 1951. The uptake of vanadium by Tunicates. *Biol. Bull.* 101:84–94.

Goodbody, I. 1957. Nitrogen excretion in *Ascidiacea.* I. Excretion of ammonia and total non-protein nitrogen. *Jour. Exp. Biol.* 34:297–305.

———. 1963. The biology of *Ascidia nigra* (Savigny). II. Development and survival of young Ascidians. *Biol. Bull.* 124:21–44.

Grave, C. 1921. *Amaroucium constellatum* (Verrill). II. The structure and organization of the tadpole larva. *Jour. Morph.* 36:71–101.

———. 1926. *Molgula citrina* (Alder and Hancock). Activities and structure of the free-swimming larva. *Jour. Morph.* 42:453–71.

———. 1932. The Botryllus type of Ascidian larva. *Papers Tortugas Lab.* 28:143–56.

———. 1944. The larva of *Styela (Cynthia) partita*; Structure, activities and duration of life. *Jour. Morph.* 75:171–73.

Hartmeyer, R. 1923–24. Ascidiacea. I. and II. Zugleich eine Übersicht über die arktische und boreale Ascidienfauna auf tiergeographischer Grundlage. *Danish Ingolf-Expedition II*, part 6:1–365; part 7:1–275.

Hattori, H., and T. Tokioka. 1953. *Ascidians of Sagami Bay. Collected by His Majesty the Emperor of Japan.* Tokyo: Iwanami Shoten. Japanese part, pp. 1–164; English part, pp. 165–307.

Herdman, W. A. 1882–1886. Report on the Tunicata collected during the voyage of the H. M. S. Challenger 1873–1876. Edinburgh. Part 1, Ascidiae simplices, vol. 6; Part 2, Ascidiae compositae, vol. 4; Part 3, Pelagic Tunicata and appendix to Part 1., vol. 27.

Huntsman, A. G. 1921. Age determination, growth, and symmetry in the test of the ascidian *Chelyosoma. Trans. Royal Canadian Inst. Toronto* 13:27–38.

Hutchinson, G. E. 1961. The biologist poses some problems. *Oceanography.* Amer. Assoc. Adv. Sci. Publ. 67: 85–94.

Huus, J. 1937. *Tunicata: Ascidiacea.* In Kükenthal. W. *Handbuch der Zoologie.* Berlin, vol. 5, 2nd half, pp. 545–672.

Jaekel, O. 1915. Über fragliche Tunicaten aus dem Perm siciliens. *Paleont. Zeitschr.* 2:66–74.

Joyce, E. A., and Jean Williams. 1969. Memoires of the Hourglass Cruises. Vol. 1, part c, Rationale and Pertinent Data. Marine Research Laboratory, Florida Dept. of Marine Resources.

Komai, T. 1951. The homology of the notochord in pterobranchs and enteropneusts. *Amer. Nat.* 85:270.

Kott, Patricia. 1969. *Antarctic Ascidiacea.* Volume 13, Antarctic Research Series. American Geophysical Union. Nat'l Acad. Sciences Publ. No. 1725.

Kowalewsky, A. O. 1867. Entwicklungsgeschichte der einfachen Ascidien. *Mem. Acad. Sci. St. Petersb.* 10 (7): 1–19.

Kupffer, C. W. von 1870. Die Stammverwandschaft zwischen Ascidien und Wirbelthieren. *Arch. Mikr. Anat.* 5, 6:115–72.

Linnaeus, C. 1767. *Systema naturae.* 3 volumes. Stockholm (Ascidiacea in pt. 1 only).

Lohmann, H. 1933. *Tunicata: Thaliacea.* In W. Kükenthal, *Handbuch der Zoologie*, vol. 5, 2nd half, pp. 1–202.

Meglitsch, P. A. 1967. *Invertebrate Zoology.* London: Oxford University Press, pp. 1–961. (The most accurate and useful of recent general volumes.)

Metcalf, M. M. 1895. Notes on tunicate morphology. *Anat. Anz.* 11:329–40.

Millar, R. H. 1959. *Ascidiacea.* Scientific results Danish Deepsea Expedition around the world 1950–52. *Galathea* Rept. 1, pp. 189–209.

Milkman, J. D., and K. O. Emery, 1968. Sea levels during the past 35,000 years. *Science* 162:1121–23.

Monniot, Claude, et Francoise Monniot. 1961. Recherches sur les Ascidies interstitielles des gravelles à Amphioxus. *Vie et Milieu*, tome 12, fasc. 2.

———. 1963. Presence à Bergen et Roscoff de Pyuridae Psammicoles du genre Heterostigma. Sarsia 13.

Monniot, C. 1965. Étude systématique et évolution de la famille des *Pyuridae* (*Ascidiacea*). *Mémoires du Muséum National d'Histoire Naturelle*, ser. A, tome 36.

Monniot, F. 1965. Ascidies interstitielles des côtes d'Europe. *Mémoires du Muséum National d'Histoire Naturelle*, ser. A, tome 35.

Monniot, C. 1969. Les Molgulidae des Mers Européenes. *Mémoires du Muséum National d'Histoire Naturelle*, ser. A, tome 60, Fasc. 4.

Monniot, F. 1970. *Cystodytes incrassatus* n. sp. Ascidie fossile du Pliocène breton. *C. R. Acad. Sci. Paris* 271: 2380–82.

Monniot, C., et F. Monniot. 1970. Les Ascidies des grandes profundeurs recoltées par les navires Atlantis, Atlantis II, et Chain. *Deep Sea Research* 17:317–36. Pergamon Press.

Plough, H. H., and N. Jones. 1940. *Ecteinascidia tortugensis* sp. nov., with a review of the Perophoridae. Carnegie Inst. Wash. Publ. No. 517:47–60.

Plough, H. H. 1969. Genetic polymorphism in a stalked Ascidian from the Gulf of Maine. *Jour. Heredity* 60: 193–205.

Romer, A. S. 1958. The early evolution of fishes. *Quart. Rev. Biol.* 21:33–69.

Salfi, M. 1933. Osservazioni sulla evoluzione delle colonie e sulla sviluppo degli abbozzi blastogenetici dei Didemnidae. *Arch. Zool. Ital.* 18:203.

Scourfield, D. J. 1937. An anomalous fossil organism possibly a new type of *Chordata*, from Upper Silurian of Sesmahagow, Lanarkshire—*Ainiktozoon. Proc. Roy. Soc.* (London) B121:533–47.

Stensio, E. 1968. *The Cyclostomes with Special Reference to the Diphyletic Origin of* PETROMYZONTIDAE *and* MYXINOIDEA. Stockholm: Almquist and Wiksen.

Sumner, F. B., R. C. Osburn, and L. J. Cole. 1913. A biological survey of the waters of Woods Hole and vicinity. *Bull. U. S. Bur. Fisheries.* 31:1–86.

Van Name, W. G. 1912. Simple ascidians of the coasts of New England and the neighboring British provinces. *Proc. Boston Soc. Nat. Hist.* 34:439–619.

———. 1921. Ascidians of the West Indian region and southeastern United States. *Bull. Amer. Mus. Nat. Hist.* 44:283–494.

———. 1930. The ascidians of Porto Rico and the Virgin Islands. Scientific Survey Porto Rico and Virgin Islds. *New York Acad. Sci.* 10:401–512.

———. 1945. The North and South American Ascidians. *Bull. Amer. Museum Nat. Hist. N.Y.* 84:1–476.

Verrill, A. E., 1871 and ff. Descriptions of some imperfectly known and new Ascidians from New England. *Amer. Jour. Science*, ser. 3, vol. 1:54–58, 93–100, 288–94, 443–46.

Waterman, A. J., and many associates. 1971. *Chordate Structure and Function.* New York: Macmillan Company, London: Collier–Macmillan, chapter 2. A Framework for further study of chordate animals, pp. 42–90.

Whiteaves, J. F. 1900. Catalogue of the marine Invertebrata of eastern Canada. *Geol. Survey Canada.* 722:1–272.

Wigley, R. L. 1961. Bottom sediments of Georges Bank. *Jour. Sedimentary Petrology* (June 1961):165–88.

Index

Acorn worm, 103
Africa, 90, 99
Agnesia glacialis, 18(Table I), 19
Agnesiidae, 18(Table I), 19, 55(Table III)
Ainiktozoon, ix, 99
Albatross IV cruise, 5, 7(incl. Fig. 4), 8
Algae, 21, 65
Amaroucium, 63
American Museum of Natural History, x
Amphioxus, ix, 103, 106(Fig. 52c)
Antarctic, 19
Apalachee Bay, Fla., x, 9, 10(Fig. 5), 12, 14(Fig. 8), 15, 58, 59, 60, 67, 89, 91
Aplidium, 48, 51, 63
Aplidium bermudae, 13(Fig. 7), 14, 15 (Fig. 10), 55(Table III)
Aplidium constellatum, 5, 10(Fig. 5), 13, 17(Table I), 20, 27(Plate V), 28(Plate VI), 55(Table III), 64, 95(Table IV); tadpole, 105, 106(Fig. 51a)
Aplidium exile, 10(Fig. 5), 13(Fig. 7), 14(incl. Fig. 8b), 55(Table III), 64(Fig. 28e), 65
Aplidium glabrum, 13, 15(Fig. 10), 48 (incl. Table II), 49(Fig. 22), 55(Table III), 63, 64(Fig. 28c), 99
Aplidium pallidum, 13, 17(Table I), 30 (Plate VIII), 48, 55(Table III), 63(incl. Fig. 28a), 64(Fig. 28b), 95(Table IV), 101
Aplidium pellucidum, 13, 55(Table III), 64(incl. Fig. 28d)
Aplidium stellatum, 13, 55(Table III), 64(incl. Fig. 28f)
Aplousobranchia, 16–18(incl. Table I), 20, 45, 54(Table III), 57, 95(Table IV)
Ascidia, 29, 51, 73
Ascidia callosa, 5, 6(Fig. 3), 13, 48(Table II), 49(Fig. 22), 55(Table III), 72(incl. Fig. 32b,i)
Ascidia corelloides, 55(Table III), 69, 72(Fig. 32g)
Ascidia curvata, 55(Table III), 72(Fig. 32f), 73
Ascidia interrupta, 55(Table III), 72(Fig. 32d), 73
Ascidia mentula tadpole, 73, 104, 105(Fig. 48)
Ascidia nigra, 10(Fig. 5), 14, 15(Fig. 10), 21, 55(Table III), 72(Fig. 32e,k), 73
Ascidia obliqua, 13, 55(Table III), 72(incl. Fig. 32c,j), 99, 100
Ascidia prunum, 5, 10(Fig. 5), 13, 15(Fig. 10), 18(Table I), 21, 40(Fig. 15), 55(Table III), 71, 72(Fig. 32a,h), 95(Table IV), 101
Ascidiidae, 18(Table I), 21, 48(Table II), 52, 55(Table III), 71, 95(Table IV)
Asia, 73, 97(incl. Table V)
Atrial siphon, 1(incl. Fig. 1), 22
Australia, 58, 60, 76, 89
Azores, 76, 78, 98

Bahamas, 68
Bay of Fundy, 5, 63, 73, 82, 99
Bdellostoma stouti, 107(incl. Figs. 53 & 54), 108(Fig. 55)
Beaufort, N.C., 10(Fig. 5), 12, 14, 54, 57, 65, 75
Bering Sea, 66, 101
Bermuda, 58, 60, 65, 66, 68, 75, 89, 90
Bipennaria, 103, 104(Fig. 46b)
Block Island, R.I., 51, 63, 86, 87
Blood flow reversal, 1
Boltenia, ix, 51
Boltenia echinata, 13, 18(Table I), 22, 31(Plate IX), 56(Table III), 74(Fig. 33c,d), 75, 96(Table IV), 100–02
Boltenia ovifera, frontispiece, 4, 6(Fig. 3), 7, 13, 18(Table I), 22, 48(Table II), 49(incl. Fig. 22), 56(Table III), 73, 74 (Fig. 33a,b), 99, 100
Bolteniinae, 18(Table I), 48(Table II), 49, 56(Table III), 73, 96(Table IV)
Boston Bay, 58, 60
Bostrichobranchus, 51, 85
Bostrichobranchus pilularis, 5, 10(Fig. 5), 14, 18(Table I), 22, 35(Plate XIII), 48(Table II), 50(incl. Fig. 23), 56 (Table III), 84, 96(Table IV)
Botryllinae, 18(Table I), 19, 22, 56(Table III), 86–88, 96(Table IV)
Botrylloides aureum, 56(Table III), 88 (Fig. 43c), 89
Botrylloides nigrum, 56(Table III), 88 (Fig. 43d), 89
Botryllus, 19, 22
Botryllus planus, 56(Table III), 88(Fig. 43b), 89
Botryllus schlosseri, 14(incl. Fig. 8b), 18(Table I), 36(Plate XIV), 56(Table III), 88(incl. Fig. 43a), 96(Table IV)
Bottom, sea, 4, 13
Branchial sac, 1, 2, 5, 19, 47–53
Branchiostoma, ix
Browns Bank, 73
Budding, 2 (Fig. 2), 3, 16, 20, 21, 39, 65, 94
Buzzards Bay, Mass., x, 65, 81

California, 66, 82
Cambrian, 103, 107
Cape Ann, Mass., 12, 14, 21, 69, 74, 86, 87, 91, 101
Cape Cod, Mass., 5, 7, 12, 22, 47, 53, 57, 60, 64, 66, 76, 78, 79, 81, 82, 84, 90, 91, 98, 100, 101
Cape Hatteras, N.C., x, 53, 58, 67, 73, 81, 84, 92
Cape May, N.J., 78, 84, 90
Carolina, 60
Casco, Bay, Me., 65, 82, 89
Cephalochordata, ix, 103
Chaetognatha, 103
Chatham, Cape Cod, 4, 77
Chelyosoma macleayanum, 10(Fig. 5), 13, 15(Fig. 10), 17(Table I), 39(Fig. 14), 55(Table III), 69, 70(Fig. 31c), 71(Fig. 31d), 95(Table IV), 101
Chordata, ix, 103
Ciona, ix, 19–21, 45, 48, 51, 57, 86; tadpole 104, 105(Fig. 49a)
Ciona intestinalis, 2, 10(Fig. 5), 13, 15 (Fig. 10), 17(Table I), 23(Plate I), 48(Table II), 49(Fig. 22), 54(Table III), 57(incl. Fig. 24), 95(Table IV)
Ciona tenella, 58
Cionidae, 17(Table I), 20, 21, 48(Table II), 52, 54(Table III), 57
Clavelina gigantea, 13(Fig. 7), 14, 20, 24(Plate II), 54(Table III), 58(incl. Fig. 25)
Clavelina oblonga, 13(Fig. 7), 14(incl. Fig. 8b), 15(Fig. 10), 17(Table I), 20,

24(Plate II), 54(Table III), 58(incl. Fig. 25), 95(Table IV)
Clavelina picta, 10(Fig. 5), 14, 54(Table III), 58(incl. Fig. 25)
Clavelinidae, 17(Table I), 21, 54(Table III), 58, 95(Table IV)
Clavelininae, 17(Table I), 20, 52, 54 (Table III), 58, 95(Table IV)
Cnemidocarpa, 19, 52
Cnemidocarpa mollis, 4, 10(Fig. 5), 14, 18(Table I), 39, 44(Fig. 19), 48(Table II), 50(Fig. 23), 51, 56(Table III), 87(Fig. 42d), 90, 96(Table IV), 101
Cnemidocarpa mortenseni, 44(Fig. 19), 56(Table III), 87(Fig. 42f), 91
Cnemidocarpa verrucosa, 90
Continental drift, x, 84–85, 94–102
Corella, 45, 101
Corella borealis, 17(Table I), 39(Fig. 14), 55(Table III), 69, 70(Fig. 31b), 95 (Table IV)
Corellidae, 17(Table I), 19, 21, 55(Table III), 68, 95(Table IV)
Corellinae, 17(Table I), 21, 55(Table III), 68, 69, 95(Table IV)
Corpus Christi, Tex., ix, 22
Crab Ledge, Cape Cod, 7, 77
Cratostigma singulare, 13, 18(Table I), 22, 32(Plate X), 56(Table III), 77(incl. Fig. 34e), 78(incl. Fig. 35), 96(Table IV), 98
Cretaceous, 97(Table V), 99, 101
Cyclostomata, 107
Cystodytes, ix, 102
Cystodytes dellechiajei, 55(Table III), 59, 61(Fig. 26f), 95(Table IV)

Davis Strait, 1, 58, 63, 75, 97
Delaware, x, 90
Dendrodoa, 19, 51, 86
Dendrodoa aggregata, 86
Dendrodoa carnea, 10(Fig. 5), 14, 15(Fig. 10), 18(Table I), 39, 45(Fig. 20), 51, 56(Table III), 86, 92(Fig. 44g), 101
Dendrodoa pulchella, 56(Table III), 86, 92(Fig. 44f), 96(Table IV)
Depth, 4, 47, 48, 54–56(Table III), 90, 94
Diazona, 19
Diazona violacea, 17(Table I)
Diazonidae, 17(Table I), 19, 54(Table III)
Dicarpa simplex, 18(Table I), 45, 46(Fig. 21), 48(Table II), 50(Fig. 23), 51, 52, 56(Table III), 88(Fig. 43f), 92, 94
Didemnidae, 17(Table I), 20, 21, 52, 55 (Table III), 65, 95(Table IV)
Didemnum, 65; tadpole, 9
Didemnum albidum, 5, 6(Fig. 3), 13, 17(Table I), 20, 29(Plate VII), 55 (Table III), 65, 66(Fig. 29c), 95 (Table IV)
Didemnum amethysteum, 13(Fig. 7), 55 (Table III), 67
Didemnum candidum, 10(Fig. 5), 13(Fig. 7), 14(incl. Fig. 8b), 17(Table I), 20, 29(Plate VII), 55(Table III), 66(incl. Fig. 29b), 95(Table IV)
Didemnum vanderhorsti, 55(Table III), 66(incl. Fig. 29c)
Diplosoma, 65
Diplosoma macdonaldi, 17(Table I), 20, 30(Plate VIII), 55(Table III), 66(Fig. 29g), 67, 95(Table IV)
Distaplia, 48, 51, 60; tadpole, 9
Distaplia bermudensis, 10(Fig. 5), 13(Fig. 7), 14(incl. Fig. 8b), 15(Fig. 10), 17 (Table I), 20, 25(Plate III), 26(Plate IV), 48(incl. Table II), 49(Fig. 22), 55(Table III), 60, 62(Fig. 27b), 101
Distaplia clavata, 13, 55(Table III), 60, 62(Fig. 27a,d), 95(Table IV), 102
Distaplia cylindrica, 48
Distaplia stylifera, 55(Table III), 60, 62 (Fig. 27c), 95(Table IV)
Doliolum, ix, 4
Dorsal tubercle, 1(incl. Fig. 1), 2
Duke University Marine Laboratory, 12, 51, 92

Eastport, Me., 60, 63, 67, 75, 82, 86, 99, 101
Echinoclinum verrilli, 13(Fig. 7), 55(Table III), 66(Fig. 29h), 67
Echinodermata, 103, 104(Fig. 46a,b)
Ecteinascidia conklini, 55(Table III), 68
Ecteinascidia tortugensis, ix, 1(Fig. 1), 2(incl. Fig. 2), 17(Table I), 22(Fig. 13), 55(Table III), 68(incl. Fig. 30d)
Ecteinascidia turbinata, 10(Fig. 5), 14(Fig. 8b), 17(Table I), 21, 22(Fig. 13), 55(Table III), 68(incl. Fig. 30e), 95 (Table IV), 101
Endostyle, 1(Fig. 1), 2, 48
Enterogona, 16–18(incl. Table I), 45, 52, 54(Table III), 57, 67, 94, 95(Table IV)
Enteropneusta, 103, 104(Fig. 46c)
Eocene, 97(Table V), 98
Eudistoma, 58
Eudistoma capsulatum, 13(Fig. 7), 55 (Table III), 58, 60(Fig. 26b,c)
Eudistoma carolinense, 55(Table III), 59, 61(Fig. 26e)
Eudistoma hepaticum, 13(Fig. 7), 55 (Table III), 59, 60(Fig. 26d)
Eudistoma olivaceum, 10(Fig. 5), 14(Fig. 8b), 17(Table I), 20(incl. Fig. 11), 55(Table III), 58, 59(Fig. 26a), 95 (Table IV), 101
Eudistoma tarponense, 55(Table III), 59
Eugyra arenosa padrensis, 56(Table III), 84
Eugyrinae, 18(Table I), 48(Table II), 50, 56(Table III), 84, 96(Table IV)
Euherdmania, 19
Euherdmania claviformis, 17(Table I)
Euherdmaniinae, 17(Table I), 19, 55 (Table III)
Europe, 5, 57, 66, 68, 73, 82, 83, 97(Table IV), 98–101
Eusynstyela, 89
Evolution, 45, 47–52, 84, 85
Eye spot, 2, 20, 45

Fisheries, National Marine, x, 5, 8, 9, 69
Florida, 13–15, 21, 39, 53, 58–60, 64–68, 73, 75, 82, 84, 86, 89, 90, 98, 101
Florida Marine Research Laboratory, 12
Food consumption, 2, 47
Fossils, ix, 59, 71, 99, 102

Georges Bank, 4, 7, 12, 14, 22, 63, 67, 73, 78, 82, 86, 90, 91, 98, 100
Georgia, x, 13, 14, 21, 22, 53, 54, 58, 60, 64, 67, 90, 101
Gloucester, Mass., 65
Gonads, 1
Gondwanaland, 99
Grand Manan Island, N.S., 58
Greenland, 1, 57, 65, 69, 73, 82, 86, 89, 97–99
Gulf of Maine, x, 4–8, 10(Fig. 5), 12, 21, 22, 49, 51, 53, 57, 60, 63, 65, 67, 69, 71–75, 81, 84, 86, 88–91, 98, 99, 101
Gulf of Mexico, 12, 21, 54, 58, 67, 90, 98, 101
Gulf of St. Lawrence, 63, 82, 84, 86, 91
Gulf Stream, 1, 11, 13

Hagfish, 107
Halocynthia pyriformis, 10(Fig. 5), 13, 56(Table III), 74(Fig. 33e,f,g), 75, 96 (Table IV)
Halocynthia tadpole, 104, 105(Fig. 49b)
Heterostigminae, 18(Table I), 19, 22, 56 (Table III), 76, 96(Table IV)
Holozoinae, 17(Table I), 48(incl. Table II), 52, 55(Table III), 59, 95(Table IV)
Hourglass cruises, 12, 13(Fig. 7)
Hudson Canyon, 51, 90, 92
Hypobythiidae, 17(Table I), 19, 55(Table III)

Ice age, 98
Iceland, 19, 75, 97, 98
Indian Ocean, 19, 60, 66, 90, 93, 97(Table V), 98, 101
Infundibula, 22, 47, 48, 50(incl. Fig. 23a,f), 79, 84
Iron, 2

Jamaica, 68

Labrador, 1, 82
Larvacea, ix, 4
Lissoclinum aureum, 55(Table III), 66 (Fig. 29i), 67
Lissoclinum fragile, 55(Table III), 67
Live bottom habitat, 14, 53–54, 60, 84, 93
Long Island, N.Y., x, 10(Fig. 5), 54, 64, 65, 81, 86, 90, 91, 98, 100, 101

Manganese, 85
Marine Biological Laboratory, x, 5, 8
Martha's Vineyard, Mass., 67, 79, 81, 90

Maryland, x, 9, 10(Fig. 5)
Massachusetts, 101
Mediterranean, 57, 82, 88, 97
Megalodicopia, 19
Megalodicopia hians, 17(Table I)
Mesozoic, x, 71, 90, 94, 99, 101
Microcosmus exasperatus, 56(Table III), 76, 77(Fig. 34c)
Migration, 5, 84, 94
Molgula ampulloides, 101
Molgula arenata, 4, 10(Fig. 5), 11, 14, 18(Table I), 22, 34(Plate XII), 56 (Table III), 83, 85(Figs. 40 & 41), 94, 96(Table IV), 101
Molgula citrina, 14, 15(Fig. 10), 18(Table I), 22, 33(Plate XI), 48(Table II), 49, 50(Fig. 23a), 51(Fig. 23g), 56(Table III, 79, 80(Fig. 36d), 84, 96(Table IV), 99–101; tadpole, 104, 105(Fig. 50)
Molgula complanata, 14, 18(Table I), 22, 33(Plate XI), 56(Table III), 79, 80(Fig. 36e), 81(Fig. 37a), 84, 102; tadpole, 79, 82(Fig. 37b), 104
Molgula griffithsii, 56(Table III), 83 (incl. Fig. 39b)
Molgula habensis, 83
Molgula lutulenta, 56(Table III), 83
Molgula macrosiphonica, 82
Molgula manhattensis, 5, 10(Fig. 5), 14, 18(Table I), 22, 34(Plate XII), 53, 56 (Table III), 80(Fig. 36a), 81, 96(Table IV), 99, 100
Molgula occidentalis, 5, 10(Fig. 5), 13 (Fig. 7), 14, 15(Fig. 10), 56(Table III), 81, 82(incl. Fig. 38), 84, 96(Table IV), 101
Molgula provisionalis, 56(Table III), 80 (Fig. 36b), 82
Molgula retortiformis, 14, 18(Table I), 22, 34(Plate XII), 56(Table III), 80, 83 (incl. Fig. 39c), 85, 101
Molgula robusta, 56(Table III), 80(Fig. 36c), 81
Molgula siphonalis, 5, 6(Fig. 3), 14, 56 (Table III), 82, 83(Fig. 39a), 85, 96 (Table IV), 99
Molgulidae, 1, 18(Table I), 19, 22, 45, 49, 51, 52, 56(Table III), 78, 94, 96(Table IV)
Molgulinae, 18(Table I), 48(Table II), 49, 56(Table III), 79, 96(Table IV)
Mt. Desert Island, Me., 5, 14, 65, 86
Mucus, 1, 2, 47, 48

Nantucket Sound, x, 5, 10(Fig. 5), 74, 86, 100
Naragansett Bay, x, 92
Newfoundland, 84
New Jersey, x, 9, 10(Fig. 5), 57
Niobium, 2
Nobska, Cape Cod, 64
Norway, 75, 88
Notochord, ix, 2, 103, 104
Nova Scotia, 5, 12, 82, 91

Ocellus, 2, 103
Octacnemidae, 18(Table I), 19, 20, 55 (Table III)
Octacnemus bythius, 18(Table I)
Oikopleura, ix
Oligotrema psammites, 18(Table I), 19
Oligotreminae, 18(Table I), 19
Oral siphon, 1(incl. Fig. 1), 19, 22
Ordovician, 107
Otolith, 2, 20, 45, 47, 103
Oxygen consumption, 47

Pacific, 19, 73, 75, 83, 91, 92, 99–101
Padre Island, Tex., x, xi, 12, 15(Fig. 9), 22, 53, 58, 67, 73, 79, 84
Paleozoic, ix, 45, 52, 98, 99, 105, 107
Pelonaia corrugata, 51, 56(Table III), 86, 92(Fig. 44e)
Penobscot Bay, Me., 7, 74
Permian, 71
Perophora, 48, 51
Perophora bermudensis, 21(Fig. 12), 55(Table III), 68(incl. Fig. 30b)
Perophora listeri, 68
Perophora viridis, 14(incl. Fig. 8b), 17 (Table I), 21(incl. Fig. 12), 48(Table II), 49(Fig. 22), 55(Table III), 67, 68(Fig. 30a,c), 95(Table IV); tadpole, 105, 106(Fig. 51b)
Perophoridae, 17(Table I), 21, 48(Table II), 52, 55(Table III), 67, 95(Table IV)
Phlebobranchia, 16–18(incl. Table I), 45, 55(Table III), 67, 95(Table IV)
Photolith, 86, 89
Pleurogona, 16–18(incl. Table I), 21, 45, 52, 56(Table III), 73, 94, 96(Table IV)
Pliocene, 59, 99
Pluteus, 103, 104(Fig. 46b)
Point Judith, R.I., 64
Pollution, 81, 91
Polyandrocarpa, 19, 52
Polyandrocarpa floridana, 37(Plate XV), 56(Table III), 89
Polyandrocarpa maxima, 14(Fig. 8b), 15 (Fig. 10), 18(Table I), 37(Plate XV), 39, 48(Table II), 50(incl. Fig. 23), 56(Table III), 88(Fig. 43g), 89, 96 (Table IV)
Polyandrocarpa tincta, 56(Table III), 89
Polycarpa albatrossi, 56(Table III), 90, 96(Table IV)
Polycarpa circumarata, 14(incl. Fig. 8b), 15(Fig. 10), 18(Table I), 38 (Plate XVI, 39, 56(Table III), 86, 87(Fig. 42c), 90, 94
Polycarpa fibrosa, 5, 6(Fig. 3), 14, 18 (Table I), 19, 39, 43(Fig. 18), 51, 56 (Table III), 86, 87(Fig. 42a,e), 90, 94, 96(Table IV), 101
Polycarpa obtecta, 13(Fig. 7), 14, 38 (Plate XVI), 39, 50, 56(Table III), 86, 87(Fig. 42b), 90
Polycarpa spongiabilis, 90
Polycitorinae, 17(Table I), 20, 52, 55 (Table III), 58, 95(Table IV)
Polyclinidae, 17(Table I), 21, 48(Table II), 55(Table III), 63, 95(Table IV)
Polyclininae, 17(Table I), 20, 52, 55(Table III), 63
Polyzoinae, 18(Table I), 19, 39, 48(Table II), 56(Table III), 86, 88, 89, 96(Table IV)
Puerto Rico, 68
Pyloric budding, 65
Pyrosoma, ix
Pyura vittata, 11, 14, 18(Table I), 31 (Plate IX), 56(Table III), 75, 76(Fig. 34a,b), 77(Fig. 34d), 96(Table IV)
Pyuridae, 18(Table I), 19, 21, 45, 49, 52, 56(Table III), 73, 96(Table IV)
Pyurinae, 18(Table I), 56(Table III), 75, 96(Table IV)

Race Point, Cape Cod, 77
Red Sea, 66, 76, 89, 98
Rhabdopleura, 106(Fig. 52a)
Rhode Island, 101
Rhodosomatinae, 17(Table I), 21, 52, 55 (Table III), 68, 95(Table IV)
Rhodosoma turcicum, 69, 95(Table IV)
Rhodosoma wigleii, 17(Table I), 39(Fig. 14), 55(Table III), 69(incl. Fig. 31a)

Sable Island, N.S., 65, 86
St. Petersburg, Fla., 12
Salpa, ix, 4
Sapelo Island, Ga., 9, 10(Fig. 5), 11(incl. Fig. 6), 14, 15, 57, 60, 65, 67, 73, 75, 76, 84, 85(Fig. 41), 94
Sea stars, 103
Sea urchins, 103
Silurian, ix, 99
South America, 66, 75, 99
South Atlantic, 19, 90
Spitzbergen, 69
Statoblast, 58
Stigmata, 1(incl. Fig. 1), 22, 47–52
Stolidobranchia, 16–18(incl. Table I), 21, 56(Table III), 73, 96(Table IV)
Stolon, 3, 89
Straits of Magellan, 58
Styela, 19
Styela atlantica, 56(Table III), 91, 92(Fig. 44c)
Styela clava, 56(Table III), 92
Styela coriacea, 56(Table III), 91, 92(Fig. 44d)
Styela nordenskjoldi, 85, 91
Styela partita, 4, 5, 10(Fig. 5), 14, 18 (Table I), 39, 41(Fig. 16), 56(Table III), 85, 91, 92(Fig. 44a), 96(Table IV), 99, 101
Styela plicata, 10(Fig. 5), 13(Fig. 7), 14, 15(Fig. 10), 18(Table I), 39, 42(Fig. 17), 38(Plate XVI), 48(Table II), 50 (Fig. 23), 51, 52, 56(Table III), 91, 92(Fig. 44b), 96(Table IV), 101
Styela rustica, 91
Styelidae, 18(Table I), 19, 22, 45, 50, 51, 52, 56(Table III), 85, 94, 96(Table IV)

Styelinae, 18(Table I), 39, 45, 48(Table II), 50, 51, 56(Table III), 86, 90, 96 (Table IV)
Suez, 57
Suspension feeding, 2, 13, 47
Sweden, 76, 77
Symplegma, 19
Symplegma viride, 13(Fig. 7), 14, 18 (Table I), 37(Plate XV), 39, 56(Table III), 88(Fig. 43e), 89, 96(Table IV), 98, 102
Synoicum pulmonaria, 65

Tadpole larva, ix, 2(incl. Fig. 2), 4, 9, 20–22, 45, 49, 73, 75, 82(Fig. 32b), 84, 85, 94, 103, 104, 105(incl. Figs. 48, 49, & 50), 106(Figs. 51a,b)
Tampa Bay, Fla., x, 9, 10(Fig. 5), 12, 13(Fig. 7), 58, 60
Temperature, 4
Tethys Sea, 97(Table V), 98, 99
Thaliacea, ix, 4
Tornaria, 103, 104(Fig. 46c)
Tortugas Island, Fla., x, 10(Fig. 5), 12, 15, 53, 67, 68, 73, 75, 76, 89
Trididemnum, 65
Trididemnum orbiculatum, 13(Fig. 7), 55 (Table III), 66(Fig. 29f), 67
Trididemnum savignii, 13(Fig. 7), 14, 15(Fig. 10), 17(Table I), 20, 30(Plate VIII), 55(Table III), 66(Fig. 29e), 67
Trididemnum tenerum, 13, 55(Table III), 66(Fig. 29d), 67, 95(Table IV)
Tunicata, ix, 4
Tunicin, 1

Vanadium, 2
Vertebrata, 103
Vineyard Sound, 14, 64, 65, 82, 86
Virginia, 90

West Indies, 5, 21, 58, 60, 67–69, 73, 75, 84, 89, 91, 101
Woods Hole, Cape Cod, 5, 57, 58, 64
Woods Hole Oceanographic Institution, 9, 92

Library of Congress Cataloging in Publication Data

Plough, Harold Henry, 1892–
Sea squirts of the Atlantic continental shelf from Maine to Texas.

Includes bibliographical references and index.
1. Ascidiacea—Atlantic coast (United States). 2. Ascidiacea—Mexico, Gulf of. 3. Tunicata—Atlantic coast (United States). 4. Tunicata—Mexico, Gulf of. I. Title.

QL613.P57 596′.09′214 76–47388
ISBN 0–8018–1687–4

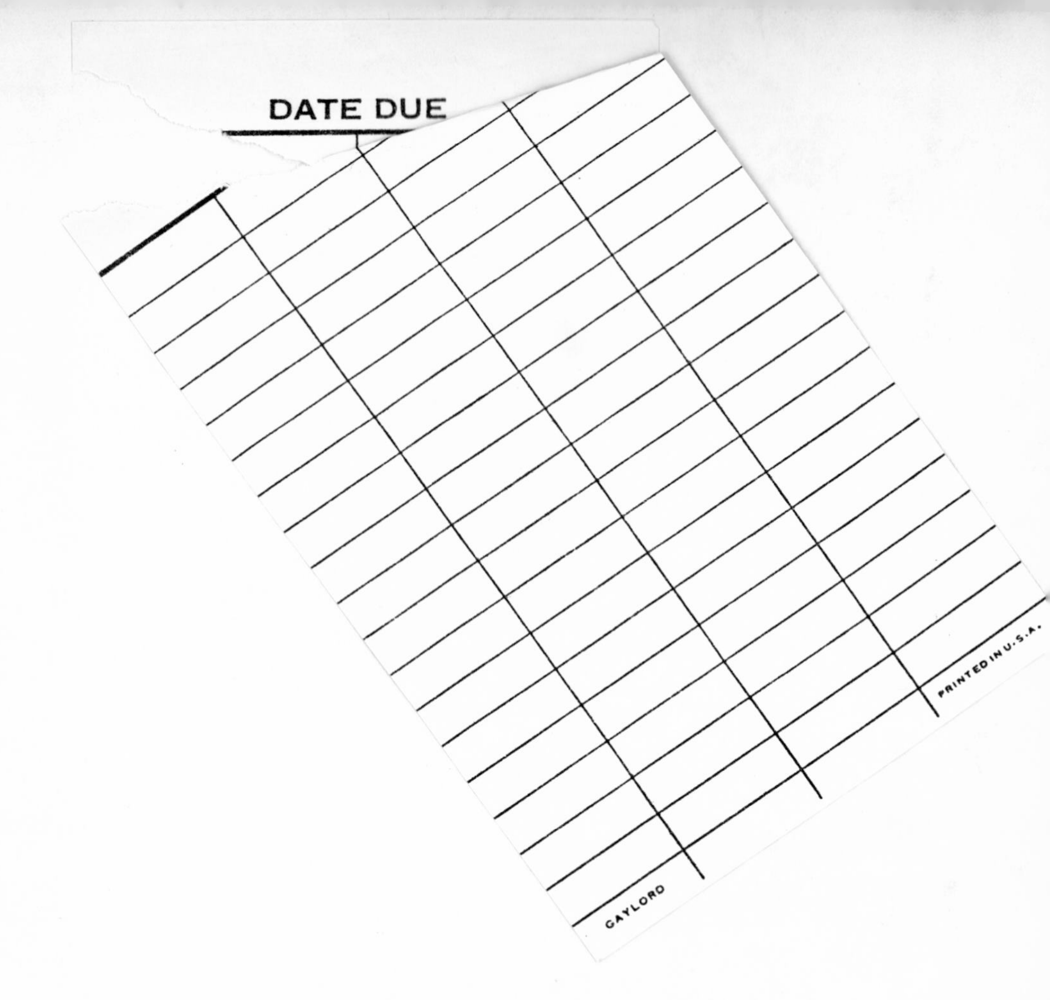